The series "Advances in Intelligent Systems and Computing" contains publications on theory, applications, and design methods of Intelligent Systems and Intelligent Computing. Virtually all disciplines such as engineering, natural sciences, computer and information science, ICT, economics, business, e-commerce, environment, healthcare, life science are covered. The list of topics spans all the areas of modern intelligent systems and computing such as: computational intelligence, soft computing including neural networks, fuzzy systems, evolutionary computing and the fusion of these paradigms, social intelligence, ambient intelligence, computational neuroscience, artificial life, virtual worlds and society, cognitive science and systems, Perception and Vision, DNA and immune based systems, self-organizing and adaptive systems, e-Learning and teaching, human-centered and human-centric computing, recommender systems, intelligent control, robotics and mechatronics including human-machine teaming, knowledge-based paradigms, learning paradigms, machine ethics, intelligent data analysis, knowledge management, intelligent agents, intelligent decision making and support, intelligent network security, trust management, interactive entertainment, Web intelligence and multimedia.

The publications within "Advances in Intelligent Systems and Computing" are primarily proceedings of important conferences, symposia and congresses. They cover significant recent developments in the field, both of a foundational and applicable character. An important characteristic feature of the series is the short publication time and world-wide distribution. This permits a rapid and broad dissemination of research results.

More information about this series at http://www.springer.com/series/11156

Piotr Kosiuczenko · Zbigniew Zieliński
Editors

Engineering Software Systems: Research and Praxis

 Springer

Editors
Piotr Kosiuczenko
Institute of Information Systems
Military University of Technology
 in Warsaw
Warsaw, Poland

Zbigniew Zieliński
Institute of Teleinformatics and Automation
Military University of Technology
 in Warsaw
Warsaw, Poland

ISSN 2194-5357   ISSN 2194-5365 (electronic)
Advances in Intelligent Systems and Computing
ISBN 978-3-319-99616-5   ISBN 978-3-319-99617-2 (eBook)
https://doi.org/10.1007/978-3-319-99617-2

Library of Congress Control Number: 2018952055

This Springer imprint is published by the registered company Springer Nature Switzerland AG
The registered company address is: Gewerbestrasse 11, 6330 Cham, Switzerland

Preface

In the last two decades, software became ubiquitous and the software systems more and more complex. This necessitates the development of more adequate software process models, more effective software engineering methods and tools supporting software development.

This book reports on some new approaches and concepts aimed at problems faced in the software development. It covers a range of topics such as systematic methods, graphical and formal models, requirements specification and validation, security, performance analysis, maintenance, real-time aspects, quality measurement. It is divided into four main parts: requirement engineering, software modelling and construction, system monitoring and performance and empirical software engineering.

The first part of this book devoted to requirements is composed of three chapters. The first chapter reports on a survey conducted in Polish IT industry and aimed at identifying most widespread challenges related to requirements. It treats the frequency of occurrence of a priori known problems in specified contexts such as the use of agile methods and smaller/larger development teams. The second chapter analyses requirement engineering in terms of monetary value and in terms related to the requirements engineering. It investigates the current state of value-based RE and the challenges that effect organizations in integrating value-based approach in the RE process. The third chapter analyses practical aspects of the use-case logic patterns approach in industrial projects. An ad hoc approach and a systematic pattern based one are compared showing an improvement in clarity, repeatability and correctness, regardless of a tool environment used.

The second part discusses issues related to software specification, design and construction. The first chapter is devoted to the problem of query definition in languages like UML and OCL. The definition prohibits any change to the system state by a query which is not a realistic. It is shown how to define queries in more general terms allowing restricted state changes. In the next chapter, an approach to multi-level security systems verification based on Bell-La Padula and Biba models is presented. It is based on models' integration, evaluation and simulation. Properties of security policies are expressed in OCL, and a corresponding verification method is outlined. The third chapter focuses on the domain models, their proper contents, the minimization of modelling effort and the maximization of the potential benefits. These issues are addressed by a unified domain model, and its usefulness for the application in real projects is studied. The agile approach to software development provided an alternative for the heavyweight traditional methods. The fourth chapter addresses its drawbacks and outlines a new hybrid method improving the shortcomings of the agile and heavy methods.

In the third part, we discuss issues related to performance evaluation, real-time computation, system monitoring and maintenance. The first chapter thematizes the performance analysis of complex web applications: single page applications based on virtual DOM and reactive user interfaces. The chapter describes results of performance evaluation for two alternative architectures. The results may support the choice of an appropriate architecture already at the software design phase. The next chapter is devoted to a method allowing one to compute in real-time strings similar to a given pattern based on the Levenshtein metric with the help of the Wagner–Fischer algorithm. The algorithm is massively parallelized with the use of CUDA technology. The third chapter of this part describes a smart application for a city traffic monitoring based on 5G network, RFID transponders and cloud infrastructure and services for supervisory control. In the last chapter of this part, its authors present challenges associated with the monitoring and the maintenance of a large telecom system. The system consists of multiple new and legacy services; it is constantly changing, and therefore, it has to be adapted to changing business needs. Selected challenges and potential directions for future research are listed.

The last part of this book consists of two chapters. The first one aims at presenting a methodology for the similarity determination of complex software systems. The methodology includes a software systems' similarity metrics and a procedure for its determination, as well as methods and tools for measuring the similarity. An experience-based overview of software projects' quality assessment criteria is presented in the second chapter of this part. The goal is an effective management of a portfolio of software projects and their ranking with respect to various quality criteria. This section includes an experience-based assessment of software projects' quality and of relevancy of methods used for the comparison and the aggregation.

The book editors would like to express their sincere gratitude to all authors of submissions and to the reviewers for their valuable assessments necessary in the paper selection process and in the chapter improvement.

June 2018

Piotr Kosiuczenko
Zbigniew Zieliński

Contents

Requirement Engineering

What Is Troubling IT Analysts?
A Survey Report from Poland
on Requirements-Related Problems

Aleksander Jarzębowicz$^{(\boxtimes)}$ ⓘ and Wojciech Ślesiński

Department of Software Engineering, Faculty of Electronics,
Telecommunications and Informatics,
Gdańsk University of Technology, Gdańsk, Poland
olek@eti.pg.edu.pl, slesinski.wojciech@gmail.com

Abstract. Requirements engineering and business analysis are activities considered to be important to software project success but also difficult and challenging. This paper reports on a survey conducted in Polish IT industry, aimed at identifying most widespread problems/challenges related to requirements. The survey was targeted at people performing role of analyst in commercial IT projects. The questionnaire included 64 pre-defined problems gathered from a literature review and a workshop involving a small group of analysts. It was completed by 55 respondents, each of whom assessed the frequency of occurrence for pre-defined problems and optionally could report additional problems based on their work experience. A ranking of most frequent problems is presented in this paper. Additional analyses for more specific contexts: agile projects and smaller/larger development teams are also provided. Final sections of the paper include comparison of our results and results of reported surveys conducted in other countries, followed by a discussion.

Keywords: Requirements engineering · Business analysis · Survey
Problems · Challenges · Analyst

1 Introduction

Requirements engineering (RE) is a part of software development process, which focuses on interaction with stakeholders and aims at defining and maintaining system/software requirements [1]. Another commonly used term is business analysis (BA), defined as a practice of enabling change in an enterprise by defining needs and recommending solutions that deliver value to stakeholders [2], which in case of software projects can be considered a wider area that encompasses RE. RE and BA are commonly regarded as important, but also difficult activities. As many software projects end up failed or challenged, causes contributing to such outcome are analyzed. Several analyses [3–6] reveal that problems related to RE/BA e.g. incomplete requirements, lack of user involvement or unrealistic goals/expectations are among main factors leading to project failures and other difficulties.

Dependencies between correctness and efficiency of requirements-related processes and software project outcomes are also confirmed by dedicated empirical research

© Springer Nature Switzerland AG 2019
P. Kosiuczenko and Z. Zieliński (Eds.): KKIO 2018, AISC 830, pp. 3–19, 2019.
https://doi.org/10.1007/978-3-319-99617-2_1

studies e.g. influence of RE techniques, good practices and resources spent on outcomes such as stakeholders' satisfaction, quality of RE products and predictability of RE process [7], increase in productivity and quality observed as result of introducing improvements to RE processes [8], correlation between RE process maturity and project success factors (scope, schedule, budget and stakeholders' perception) [9].

All of above imply that RE/BA is an important topic which still needs new solutions and evaluation of their effectiveness. However, in order to provide a relevant solution, one needs to correctly identify the problem, which requires knowing the real picture of RE/BA in the IT industry. Such knowledge cannot be established on the basis of theoretical considerations, but has to be gathered from industry professionals involved in real-life software projects. In our case, if we intend our research to be useful to practitioners, then the most likely first recipient is the domestic IT industry.

Our aim was therefore to identify the most widespread and frequent problems affecting RE/BA activities encountered by analysts employed in Polish IT companies. At first glance, such problems are quite well known and can be found in virtually any book or course material on RE/BA. However, problems reported there are usually based on author's experience, rather than collected in systematic, scientific manner through surveys or field studies, which does not allow to generalize such results. When it comes to scientific papers on RE/BA problems, only a few sources can be found and not a single one of them concerns Polish industry. This is the identified research gap we intended to fill.

In this paper we describe a questionnaire-based survey study targeting software project analysts (i.e. people responsible for RE/BA activities). The questionnaire included a list of 64 pre-defined problems collected from a literature review and a workshop involving a small group of analysts. We gathered answers from 55 respondents, who evaluated how frequently they had encountered particular problems in their professional experience and optionally could report additional problems.

The main contributions are: the overall ranking of problems (as reported by all respondents) and separate rankings for agile project and for smaller/larger teams. The additional contributions are: review of problems reported in literature and comparison of survey outcome to similar studies from other countries.

The rest of the paper is structured as follows. In Sect. 2 we outline related work. Section 3 describes an overview of research process and its particular steps. In Sect. 4 survey's main results are presented, followed by a comparison to results obtained by others in Sect. 5. In Sect. 6 we discuss validity threats before concluding the paper in Sect. 7.

2   Related Work

We narrow down this related work summary to survey-based papers on gathering and analyzing information about RE/BA problems. Of course, many more research reports on RE/BA state of practice are available, but since they focus on RE/BA practices, techniques, process maturity etc. we consider them not to be directly related.

Several works describing research on RE/BA problems, conducted in various countries are available. Hall et al. [10] performed a case study collecting RE problems experiences in 12 UK software companies by assembling employees into focus groups and interviewing them. Solemon et al. [11] surveyed industry practitioners from Malaysia to identify most common RE problems and cross-referenced them with process maturity and good practices applied by IT companies. Liu et al. [12] conducted an industrial survey conducted in China and (among other findings) reported major failure reasons in RE practices. An ongoing research (parallel to our work) known under the name of NaPiRE (Naming the Pain in Requirements Engineering) initiative is conducting a family of replicated surveys on RE problems, their causes and consequences [13]. NaPiRE surveys have already included 10 countries and more are expected in further replications [14].

We provide more information on the abovementioned studies' findings in Sect. 5, where we compare them to our results. Nevertheless, none of those studies concerned Polish industry, nor even any other country from Central and Eastern Europe (with the exception of Estonia which is included in NaPiRE, but no results have been published yet).

Our main interest was however Polish IT industry and information about RE/BA problems in this context is very scarce. A literature search revealed no directly related work. Some industrial survey reports on software project outcomes and/or problems are available, but RE/BA issues are hardly included within their scopes. For example, a report summarizing a survey based on 80 software projects [15] shows proportions of successful, challenged and failed project. It also lists some contributing factors (e.g. project size, development methodology, risk management), but RE/BA processes and issues were not included in survey questionnaire. Another survey [16] identified several problems plaguing software development in Poland, but it does not distinguish any explicit category of RE/BA problems, only very few such problems are included and assigned to Management category. Papers dealing with particular RE/BA problems e.g. difficulties in understanding and communicating customers' needs [17] or neglecting non-functional requirements [18] can be found, but their purpose is to propose solution to a selected problem, not to analyze the broader scope of problems and their occurrences in the industry.

Also, a previous work of one of us [19, 20] should be mentioned. One part of this work was a survey on RE/BA problems, which was however a small-scale study (8 interviewees from 2 companies) and as such had its limitations. Moreover, during the interviews, only open questions were used, encouraging the respondents to enumerate RE/BA problems affecting their work. The subsequent analysis of results showed a relatively low similarity of reported problems, even in case of interviewees employed by the same company. It can indicate that they had difficulties reminding all relevant problems without any guidance provided. Such observations resulted in the idea for the research described in this paper – to conduct a survey focusing solely on RE/BA problems, involving a significantly larger and more diverse group of respondents and to provide them with a list of pre-defined problems together with the opportunity to report additional ones.

3   Research Process

After identifying the existing research gap and the need for establishing the state of practice in Polish IT, we planned the research process to be followed. We set out to answer he following research questions (RQ):

- RQ1: What are the most common requirements-related problems in Polish IT industry?
- RQ2: What are the differences in reported problems with respect to different software project contexts?

As for RQ2, by context we meant e.g. size of development team or software development approach. We suspected that such factors can make particular problems more or less frequent (e.g. agile projects can cope better with requirements changes, larger teams can have more issues about communicating requirements between developers etc.). No assumptions however were made, we only planned to include context questions in the questionnaire.

The central term used in our research is "problem", which requires some explanation. We consider RE/BA problem to be any requirements-related issue that is perceived difficult or error-prone by people involved in RE/BA processes. Some of such problems can be considered as an inevitable part of analyst's job – it is for example natural that developers and business stakeholders have difficulties understanding each other or that requirements change to reflect business domain dynamics. However, if a given issue is reported as problematic by practitioners, it is something that requires further attention and should be addressed by e.g. dedicated techniques, good practices or tool support. This is the reason we do not exclude any potential problems on the grounds of their origin or responsible party. It is consistent with the notion of "problems" or "challenges" used by others (as outlined in related work summary).

The research process we planned included the following steps (described in the following sections):

1. Literature review aimed at extracting and cross-checking RE/BA problems discussed in reviewed sources.
2. Additional identification of RE/BA problems by organizing a workshop with a group of analysts.
3. Questionnaire design.
4. Conducting the survey and analyzing results.

3.1   Literature Review

The literature search and review was aimed at gathering problems already described in sources of various types. It was not a systematic literature review (SLR), as it was not our intent to identify all possible sources and to list every problem described in the literature. Instead, we wished to cross-reference a number of pre-selected sources and use the subset of most common problems as an input to subsequent steps of our research process.

We selected 9 sources, trying to cover various forms of publications: scientific papers, books and technical reports. It was also different from the usual manner of literature review, where only peer-reviewed sources are included. We deliberately reached for other sources, even those associated with commercial software tools. The reason was to use information originating from industry. Industry professionals are likely to use forms of publications other than scientific papers e.g. informal articles, books or technical reports.

Sources S1–S4 are scientific papers. S1 [10] and S2 [11] are survey-based studies described in related work summary. The remaining papers use other approaches: S3 [21] is a literature review, which summarizes and categorizes problems found in numerous sources, while S4 [22] is a rather subjective, experience-based discussion of selected problems. Sources S5 [23] and S6 [24] are widely known books on RE, while S7 [25] is a relatively new book item published by Polish authors, which we considered to be closer to Polish IT state of practice. Sources S8 [26] and S9 [27] are technical documents associated with IT tools supporting RE/BA processes.

Table 1 summarizes most essential findings of literature review. It lists problems mentioned by most sources (at least 4 out of 9). For each problem, particular sources mentioning it and total number of them are given. Problem names were unified, as sources use different wordings. Table 1 can also support an argument that RE/BA problems are not well evidenced yet, as sets of problems mentioned by particular sources are only partially overlapping.

Table 1. Summary of RE/BA problems reported in literature.

Problem	S1	S2	S3	S4	S5	S6	S7	S8	S9	#
Incomplete requirements		X	X	X	X	X	X		X	7
Ambiguous requirements		X		X	X	X	X	X	X	7
Analysts lack adequate training/competencies	X	X	X	X		X				5
Inadequate requirements management procedures/tools			X	X	X	X			X	5
Lack of stakeholders' commitment		X	X		X	X			X	5
Changing requirements		X	X		X	X		X		5
Inconsistent requirements		X	X	X	X				X	5
Requirements defined in technical jargon rather than customer language	X		X	X					X	4
Requirements lack meta-data (source, priority, status, etc.)				X		X	X	X		4
"Obvious" requirements not reported			X	X		X		X		4

(*continued*)

Table 1. (*continued*)

Problem	S1	S2	S3	S4	S5	S6	S7	S8	S9	#
Scope creep				X	X	X		X		4
Lack of quality control of requirements		X		X		X	X			4
Inadequate tool support				X		X	X		X	4
Communication problems between customers and developers	X		X			X			X	4

3.2 Workshop

The reason of the workshop was to identify additional problems that may be more specific to Polish industry. The workshop was conducted in September 2016 and included 6 active participants: 5 analysts and 1 researcher acting as a moderator. Invited analysts represented different companies and application domains. Also their experience varied – some of them could be considered as beginners, while others quite experienced (about 7 years in RE/BA). The workshop took place in an informal setting and was planned as a moderated discussion. Prior to the workshop, literature research findings were coded by deriving common themes e.g. business goals, cooperation with business stakeholders, requirements quality, requirements prioritization, RE documentation templates etc. About 40 such themes were identified.

During the discussion, all themes were walked through. The moderator asked open questions about problems related to a given theme and participants reported such experienced and/or known problems (or lack of them). The discussion was free-form and informal, participants were also able to refer to each other's statements. The discussion was audio-recorded and later transcribed. Problems reported by participants were extracted. Most of them could be matched to those derived from literature. There were some exceptions though e.g. participants revealed situations where business goals were adjusted to (implemented) requirements, not otherwise (especially in public sector). Moreover, this discussion allowed to identify several sub-variants of already known problems e.g. "Ambiguous requirements" were divided into: "Stakeholders lack sufficient domain knowledge to define requirements", "Software Requirements Specification document is very generic" and "Specified requirements insufficiently detailed/verifiable". Such findings provided an additional input to the next step - questionnaire design.

3.3 Questionnaire-Based Survey

We chose to apply questionnaire-based survey as a method to answer research questions. We also decided to survey solely the practitioners who perform the role of analyst in software projects (i.e. who are responsible or RE/BA activities, regardless of how exactly their job positions are named). The reason was to obtain data based on first-hand experience, from people who are directly involved in RE/BA processes (as opposed to those who are only influenced by it e.g. testers). However, by selecting such profile of respondents we limited the available points of view, which could introduce bias, but it was a trade-off we considered acceptable. The introductory part of the questionnaire included a clear message that it is intended for analysts only. As our survey was targeting analysts from Poland only, the questionnaire was in Polish.

The first part of the questionnaire was supposed to collect background information about respondents. It included questions about: respondent's experience (in RE/BA, as an analyst), size of development team he/she belongs to, software development approach used in projects he/she works on (agile, plan-driven, other). Optionally, respondent could also enter his/her first name and e-mail address to receive a report on survey results afterwards.

The second part was dedicated to RE/BA problems. We included 64 problems on the basis of literature review and workshop. We selected problems that were mentioned by most of reviewed sources or reported by workshop participants. Workshop discussions also motivated us to refine some more generic problems from literature into two or more sub-variants.

We divided problems into 12 groups: business goals; project scope; sources of requirements; elicitation, analysis and specification; RE documents templates; requirements management; glossary; cooperation with stakeholders; cooperation with developers; quality assurance for requirements; analyst's competencies; other problems. The reason was to present respondents with only one group of problems at the time and thus to keep them focused. For each problem in such group, a respondent was asked to assess how frequently he/she encounters it in professional work. To answer a following 5-point Likert scale was used: 0 – "never", 1 – "rarely", 2 – "sometimes", 3 – "often", 4 – "always". For each group, an open question was also included – it was a request to report additional problems not included in pre-defined questions, but relevant to a given group.

Several iterations and reviews of questionnaire were necessary, as we paid attention to proper wording of questions, unambiguity and comprehensibility. After that a pilot study was conducted. It involved 3 people fitting the respondent's profile. All reported issues, doubts and improvement suggestions were addressed in the final version of the questionnaire. As for technical means, we checked on-line survey software services. Initially, two versions of questionnaire were prepared using Google Forms and anki-etka.pl services. During the pilot study we asked participants to compare those two versions. They perceived Google Forms version as more readable and intuitive and as such we decided to use it in our survey study.

We published the questionnaire on-line and posted invitations to participate in the survey on websites dedicated to RE/BA topics and social network groups for Polish analysts. No direct invitations to particular individuals were used. The weakness of this approach is that it does not allow to calculate the response rate and it limits our knowledge about respondents' background to the information provided in their questionnaire answers. The survey was open for 7 weeks (April 27th–June 17th 2017) and during this period we obtained 55 responses.

4 Survey Results

Context information about survey responders and their working environments is presented in Fig. 1. It is also worth mentioning that over 75% of them expressed interest in receiving a report on survey results and provided contact e-mail addresses. A complete listing of survey questions and answers (translated to English) can be found in a dataset available on-line [28].

As shown in Fig. 1, 40% of respondents had at least 5 years of experience, while 78% at least 2 years. It indicates that while senior/expert analysts were a minority, most of survey participants had sufficient experience to provide first-hand knowledge about RE/BA problems. As for team sizes, most respondents worked in teams including 6–10 members, but smaller and larger teams were represented as well. When answering the question about software development approach, 40% of respondents declared working in projects that apply agile approach, while plan-driven approach was used by only 9%. However, almost half of respondents declared that they had not followed a single approach (agile or plan-driven), but used both in different projects.

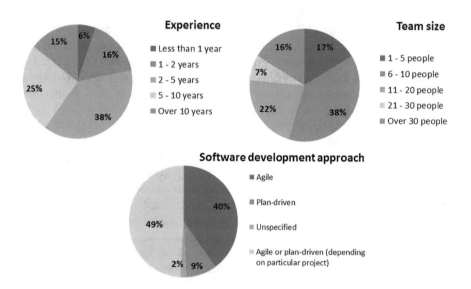

Fig. 1. Context information about respondents.

Answers to the second part of the questionnaire (assessments on frequency of occurrence of particular problems) were processed in order to create a ranking of most frequent problems. Additional (not pre-defined) problems reported by respondents were not further processed because of low similarity (no problem was reported by more than 2 people). We decided to use the mean value of answers (0–4 values, accordingly to the scale specified in Sect. 3.3) to represent frequency. A median value could be considered more appropriate for ordinal scale, but in case of 5-value scale it is unlikely to note differences.

The resulting ranking based on this metric is given in Table 2. Problem names are shortened for the sake of brevity. In the questionnaire, problems had longer descriptions and sometimes included examples in order to be well understood by respondents. The longer descriptions are available in the associated dataset [28].

Survey results clearly show that most frequent problems are related to cooperation between analysts and stakeholders, as 16 out of 20 top problems fall into such category. Among the remaining ones, P12 and P13 deal with lack of good practices applied by a

Table 2. Twenty most frequent RE/BA problems according to survey results.

ID	Problem	Never	Rarely	Sometimes	Often	Always	Mean	Std. Dev
P1	Unrealistic expectations of stakeholders	0	4	11	24	16	**2.95**	0.89
P2	Stakeholders do not express 'obvious' requirements	1	3	14	25	12	**2.80**	0.85
P3	Scope creep	0	4	14	26	11	**2.80**	0.91
P4	Too short time for analysis available	1	7	12	20	15	**2.75**	1.06
P5	Stakeholders' low availability	0	4	16	26	9	**2.73**	0.83
P6	Stakeholders describe solutions instead of requirements	0	4	17	30	4	**2.62**	0.73
P7	Stakeholders are unable to express requirements other than change requests to working software	0	6	18	25	6	**2.56**	0.83
P8	Conflicting requirements from different stakeholders	1	8	15	23	8	**2.53**	0.98
P9	Stakeholders ignore business goals and focus on requirements only	0	8	16	27	4	**2.49**	0.84
P10	Stakeholders issue requirements clearly outside project's scope	1	6	20	24	4	**2.44**	0.86
P11	Business goals are not measurable/verifiable	1	10	16	26	2	**2.33**	0.88
P12	Interdependencies between requirements are not identified/maintained	7	5	18	18	7	**2.24**	1.19
P13	No defined process for requirement changes	2	11	20	18	4	**2.20**	0.97
P14	A stakeholder believes that all requirements are essential and is unable to prioritize them	4	12	13	23	3	**2.16**	1.07
P15	A stakeholder accepts specified requirements, which he/she had not read or comprehend	3	9	22	18	3	**2.16**	0.96
P16	Specified requirements insufficiently detailed / verifiable	3	7	26	17	2	**2.15**	0.89
P17	Conflicts between stakeholders about requirements' priorities	3	10	23	16	3	**2.11**	0.96
P18	Difficult communication with a remote stakeholder	1	19	14	17	4	**2.07**	1.02
P19	Requirements are not defined by right stakeholders	2	12	22	19	0	**2.07**	1.03
P20	Stakeholders lack sufficient domain knowledge to define requirements	2	12	22	19	0	**2.05**	0.85

12   A. Jarzębowicz and W. Ślesiński

supplier (software development company or team). P4's origin is not obvious - it can be a result of customer pressure or of supplier's poor planning. P16 is probably responsibility of both sides – stakeholders issue requirements in vague form and analysts (or other development team members) do not ensure such requirements are refined. Other categories of problems e.g. cooperation between analysts and the rest of development team, analysts' competencies, documents used in RE/BA, quality management hold lower positions in the ranking - outside top 20 presented in Table 2.

To address RQ2, we also processed survey results to find out whether the same problems are reported in various contexts. One factor determining context is software development approach. We were only able to check answers for Agile, as Plan-driven population was too small (5 respondents only). Most of respondents chose answer "Agile or plan-driven, depending on particular project", which prevented us from further analysis because questionnaire design did not allow to determine which problems were associated to which approach. All other answers other that Agile (Plan-driven, Agile or Plan-driven, Unspecified) were assigned to the group "Other". The other factor examined was development team size. We decided to divide respondents w.r.t. team size

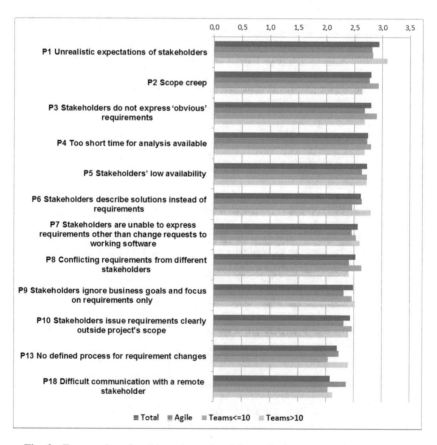

Fig. 2. Frequencies of problems in total and in particular contexts (mean values).

(≤ 10 and >10), which allowed to form two groups (30 and 25 people) and analyze answers separately. For each context (Agile, Other, Teams ≤ 10, Teams > 10) Mean values were calculated, using only answers of respondents fitting a given context. A simple comparison of mean values is shown in Fig. 2.

To verify whether differences between groups (Agile/Other and Teams $\leq$ 10/Teams > 10) are statistically significant, we used Mann-Whitney-Wilcoxon Test (suitable for ordinal scale and independent samples). Results are presented in Tables 3 and 4. Both tables include Mean values for particular groups and p-values of statistical tests for pairs assigned to each problem. Also, relative rankings of problems for each context and a ranking of total answers (from all respondents) are included in both tables (# column).

We intended to list top 10 problems from each context in Tables 3 and 4 (except "Others" which is not a particular context but a group of various ones). As can be seen, only 12 problems are sufficient do achieve it (P1–P10, P13 and P18), which indicates that similar RE/BA problems are experienced in different contexts. This impression is confirmed by statistical analysis. For this purpose, we used KNIME Analytics 3.3.1 to compute parameters and R 3.5.0 for Mann-Whitney-Wilcoxon tests (using R-Snippets in KNIME). In all cases presented in Tables 3 and 4, the p-value was greater than 0.05, which does not allow for conclusion that populations are non-identical.

There are of course some differences in mean values and relative rankings, but as they are not significant, it is quite surprising that different contexts do not introduce nor magnify specific problems (e.g. cooperation with development team for larger teams).

Table 3. Comparison of problems w.r.t. software development approach.

ID	Problem	Total	Agile		Other	p-value
		#	#	Mean	Mean	
P1	Unrealistic expectations of stakeholders	1	1	2.818	3.03	0.5166
P2	Scope creep	2	2	2.773	2.818	0.9047
P3	Stakeholders do not express 'obvious' requirements	2	4	2.682	2.879	0.3501
P4	Too short time for analysis available	4	3	2.727	2.758	0.8799
P5	Stakeholders' low availability	5	5	2.636	2.788	0.7535
P6	Stakeholders describe solutions instead of requirements	6	5	2.636	2.606	0.7308
P7	Stakeholders are unable to express requirements other than change requests to working software	7	7	2.455	2.636	0.5369
P8	Conflicting requirements from different stakeholders	8	8	2.409	2.606	0.6444
P9	Stakeholders ignore business goals and focus on requirements only	9	10	2.318	2.606	0.343
P10	Stakeholders issue requirements clearly outside project's scope	10	10	2.318	2.515	0.5927
P18	Difficult communication with a remote stakeholder	18	9	2.364	1.879	0.0827

Table 4. Comparison of problems w.r.t. team size.

ID	Problem	Total	Teams ≤ 10		Teams > 10		p-value
		#	#	Mean	#	Mean	
P1	Unrealistic expectations of stakeholders	1	3	2.833	1	3.08	0.3499
P2	Scope creep	2	1	2.933	6	2.64	0.2388
P3	Stakeholders do not express 'obvious' requirements	2	2	2.9	4	2.68	0.3532
P4	Too short time for analysis available	4	4	2.8	5	2.68	0.6594
P5	Stakeholders' low availability	5	5	2.733	3	2.72	0.8274
P6	Stakeholders describe solutions instead of requirements	6	8	2.467	2	2.8	0.0755
P7	Stakeholders are unable to express requirements other than change requests to working software	7	7	2.533	7	2.6	0.7168
P8	Conflicting requirements from different stakeholders	8	6	2.633	9	2.4	0.3362
P9	Stakeholders ignore business goals and focus on requirements only	9	9	2.467	8	2.52	0.6673
P10	Stakeholders issue requirements clearly outside project's scope	10	9	2.467	10	2.4	0.5614
P13	No defined process for requirement changes	13	19	2.033	10	2.4	0.1586

The ranking in Table 2 provides answer on RQ1 about the most common requirements-related problems in Polish industry. Tables 3 and 4 provide answer to RQ2 about differences in reported problems w.r.t. different software project contexts.

5 Comparison with Other Results

Each of sources mentioned in Sect. 2 (related work) enumerates main problems/challenges identified as result of conducted research study. Direct comparison of our results with those obtained by others is difficult because researched RE/BA problems were defined more or less differently w.r.t. names used but also to assumed abstraction levels and scope (inclusion/exclusion of issues outside RE/BA but potentially affecting that area). Despite this, we would like to compare results to such extent it is possible. When citing problems from related work, in parentheses we give the IDs of (approximately) matching problems from Table 2.

A study by Hall et al. [10] divided problems into two groups: organizational-based and process-based. For the first group there is little similarity, mainly related to "User communication" (P5, P18) and "Inadequate resources" (P4). "Company culture" is a possible match (P13(?)), but other problems ("Developer communication", "Inappropriate skills", "Staff retention" and "Lack of training") have no counterparts in our findings. More similarity can be found for process-based problems: "Vague initial requirements" (P7, P11, P16), "Poor user understanding" (P2, P20), "Requirements growth" (P3, P10), "Undefined requirements process" (P13) and "Inadequate requirements traceability" (P12). Only "Complexity of application" has no match.

Solemon et al. [11] used a similar list of problems (and division into two groups) to the one from [10], but introduced more distinctions between requirement flaws. Here we address only those additional or modified ones: "Incomplete requirements" (P14, P16, P19), "Inconsistent (changing) requirements" (P3, P8, P10), "Ambiguous requirements" (P7, P11, P16) and "Lack of defined responsibility" (P19). It is also worth mentioning that while Hall et al. claim that "our findings suggest that organizational issues exacerbate all types of requirements problems" (referring to lack of skills and staff retention as examples), Solemon et al. conclude "Our results suggest that RE problems experienced by the companies in our study can be attributed more to factors inherent within the RE process rather than to factors external to the RE process".

In case of study by Liu et al. [12], more similarities can be found. The most important problem reported by them is "Customers do not have a clear understanding of system requirements themselves, including scope of the system, major functional features and nonfunctional attributes" (P1, P7, P10, P16). For majority of other problems counterparts can also be found: "Users' needs and understanding constantly change" (P3), "Software engineers do not have access to sufficient domain knowledge and expertise" (P20), "Project schedule is too tight to allow sufficient interaction and learning period between customer and development team" (P4), "Requirements decision-makers lack of technical and domain expertise" (P19, P20), "Broken communication links between customer, analyst and developer" (P5). There are no matches for problems: "Reuse existing design in wrong context and environment" and "Lack of standardized domain data definition and system-environment interface".

Mendez Fernandez et al. [13] presented top 10 problems they found. All of them except one can be (more or less) matched to our "top 20 list" items. Below problems from [13] are listed, ordered by frequency descending: "Incomplete and/or hidden requirements" (P2, P7, P9, P14), "Communication flaws between project team and customer" (P1(?), P15), "Moving targets (changing goals, business processes and/or requirements)" (P3), "Underspecified requirements that are too abstract" (P11, P16), "Time boxing/Not enough time in general" (P4), "Communication flaws within the project team" (no match), "Stakeholders with difficulties in separating requirements from known solution designs" (P6), "Insufficient support by customer" (P5, P18, P19), "Inconsistent requirements" (P8, P12, P17) and "Weak access to customer needs and/or business information" (P20). In this case the similarity is quite high, only P10 and P13 have no counterparts and match between P1 and "Communication flaws" is questionable.

6 Validity Discussion

We are aware that our study had several limitations that can pose potential threats to validity. First of all, the study and results presented are based on reasonable but still limited number of participants. Another important issue is participants representativeness – we cannot claim that software projects and companies our respondents work for are consistent with the general picture of Polish IT. For example, plan-driven approach is poorly represented, which may stem from the widespread adoption of agile methods, but may also be a matter of this particular sample. Also, a discrepancy between declared and actual development approach is possible, as e.g. a few respondents declared agile approach and team size of 21–30 or 30+ at the same time.

We deliberately planned to survey analysts only, which has some implications. First, we cannot be sure that only analysts answered the questionnaire as our survey was not based on personal invitations. We explicitly stated target profile in published invitations and in questionnaire introductory text, but that could be ignored (however it is rather not likely that a person not involved in RE/BA would be willing to answer almost 70 questions about this topic). More important issue is that by asking analysts only, we are likely to introduce bias by limiting survey to one point of view only. An interesting observation is that problems associated with analysts' negligence or lack of competencies were among those with lowest scores. It can be the real picture of RE/BA practices and problems, but can also indicate that analysts are more likely to attribute problems to actions of other parties rather than themselves or fellow analysts.

Another threats typical to surveys are: clarity/unambiguity of questions and honesty of answers. We made a substantial effort to minimize the first threat by several reviews of the questionnaire and a pilot study. The second threat is minimized by the fact that respondents were anonymous (optionally they could provide e-mail address – but any address, not necessarily professional one). As such a respondent had no reason to hide information about problems e.g. in order to make company look better. We are also aware that our results are based on respondents' perception, not "hard data" gathered from software projects, but it is a limitation of almost any survey.

Finally, we cannot claim generalizability of results to other countries, as from start we only intended to research RE/BA problems in IT industry in Poland.

7   Conclusions

This paper identified, on the basis of industrial survey study, most frequent requirements-related problems from the point of view of IT analysts from Poland. The main contribution is the resulting list of problems, together with combined assessment metrics. The conclusion that can be drawn from this list is that the most problematic area in RE/BA is communication with stakeholders. Additional data analysis led to development of problem rankings for particular contexts (Agile development approach, smaller teams, larger teams). Such rankings show some differences, but none of them is statistically significant, thus in general the same problems are present in various contexts. Results obtained in our study were also compared to the findings of similar research studies from other countries.

Results described in this paper can be of potential value for researchers working in RE/BA area, so they can target most frequent problems by analyzing their contributing factors and by proposing and evaluating new solutions. Dedicated methods, tools and/or practices can be introduced to mitigate particular problems. Results can also be used by industry practitioners to raise awareness about problems likely to be expected and consequently to be prepared to deal with them. By practitioners we mean mostly analysts but also others e.g. project managers who are responsible for stakeholders/scope management and for planning all project activities including RE/BA. Finally, knowledge about RE/BA problems in Polish industry can be utilized in BSc/MSc requirements engineering courses and in industrial training programs intended for analysts.

Possible directions of future work include a more thorough analysis of top problems with respect to their root causes as well as identifying effective solutions to address them. Moreover, additional survey studies would be useful, especially studies involving other points of view e.g. those of developers, project managers or stakeholders representing business domain and customer's side. Also, as our survey focused on problems' frequency only, a study on problems' severity (w.r.t. consequences) would be advisable.

Acknowledgements. We would like to thank all workshop participants and survey respondents who shared with us with their knowledge and experience. Particular thanks are due to Hanna Tomaszewska from Analizait.pl for disseminating invitations to participate in the survey and to Agnieszka Landowska from Gdańsk University of Technology for advice on data analysis. We are also grateful to anonymous reviewers for their helpful suggestions.

References

1. International Organization for Standardization (ISO): ISO/IEC/IEEE 29148: Systems and Software Engineering—Life Cycle Processes—Requirements Engineering. International Organization for Standardization (ISO), Geneva (2011)
2. International Institute of Business Analysis: A guide to the business analysis body of knowledge (BABOK) 3.0 (2015)
3. The Standish Group: Chaos report (2014)
4. Arras People: Project management benchmark report (2010)

5. McManus, J., Wood-Harper, T.: Understanding the sources of information systems project failure—a study in IS project failure. Manag. Serv. **51**, 38–43 (2007)
6. Charette, R.N.: Why software fails. IEEE Spectr. **42**, 42–49 (2005)
7. Hofmann, H.F., Lehner, F.: Requirements engineering as a success factor in software projects. IEEE Softw. **18**, 58–66 (2001)
8. Damian, D., Chisan, J.: An empirical study of the complex relationships between requirements engineering processes and other processes that lead to payoffs in productivity, quality, and risk management. IEEE Trans. Softw. Eng. **32**, 433–453 (2006)
9. Ellis, K., Berry, D.M.: Quantifying the impact of requirements definition and management process maturity on project outcome in large business application development. Requir. Eng. **18**, 223–249 (2013)
10. Hall, T., Beecham, S., Rainer, A.: Requirements problems in twelve software companies: an empirical analysis. IEE Proc. Softw. **149**, 153–160 (2002)
11. Solemon, B., Sahibuddin, S., Ghani, A.: Requirements engineering problems and practices in software companies: an industrial survey. In: Communications in Computer and Information Science, CCIS, vol. 59, pp. 70–77 (2009)
12. Liu, L., Li, T., Peng, F.: Why requirements engineering fails: a survey report from China. In: Proceedings of the 18th International Requirements Engineering Conference RE 2010, pp. 317–322 (2010)
13. Mendez Fernández, D.: Naming the pain in requirements engineering: contemporary problems, causes, and effects in practice. Empir. Softw. Eng. **22**, 2298–2338 (2017). https://doi.org/10.1007/s10664-016-9451-7
14. Mendez Fernandez, D.: Supporting requirements-engineering research that industry needs: the NaPiRE initiative. IEEE Softw. **35**, 112–116 (2018)
15. Frączkowski, K., Dabiński, A., Grzesiek, M.: Raport z Polskiego Badania Projektów IT 2010 (2011). http://pmresearch.pl/wp-content/downloads/raport_pmresearchpl.pdf
16. Pieszczyk, E., Werewka, J.: Analysis of the reasons for software quality problems based on survey of persons involved in the process of developing of IT systems. Bus. Inf. **3**(37), 85–102 (2015)
17. Przybyłek, A.: A business-oriented approach to requirements elicitation. In: Proceedings of the 9th International Conference on ENASE 2014, pp. 152–163 (2014)
18. Kopczyńska, S., Nawrocki, J.: Using non-functional requirements templates for elicitation: a case study. In: Proceedings of the 2014 IEEE 4th International Work Required Patterns, RePa 2014, pp. 47–54 (2014)
19. Marciniak, P., Jarzębowicz, A.: An industrial survey on business analysis problems and solutions. In: Software Engineering: Challenges and Solutions, AISC, vol. 504, pp. 163–176. Springer (2016)
20. Jarzębowicz, A., Marciniak, P.: A survey on identifying and addressing business analysis problems. Found. Comput. Decis. Sci. **42**, 315–337 (2017)
21. Davey, B., Parker, K.R.: Requirements elicitation problems: a literature analysis. Issues Inf. Sci. Inf. Technol. **12**, 71–82 (2015)
22. Firesmith, D.: Common requirements problems, their negative consequences, and the industry best practices to help solve them. J. Object Technol. **6**, 17–33 (2007)
23. Leffingwell, D., Widrig, D.: Managing Software Requirements. Addison-Wesley, Boston (2003)
24. Wiegers, K., Beatty, J.: Software Requirements. Microsoft Press, Redmond (2013)
25. Chrabski, B., Zmitrowicz, K.: Requirements Engineering in Practice (in Polish: Inżynieria Wymagań w Praktyce). Naukowe PWN, Wyd (2015)

26. Rational Software Corporation: Using rational RequisitePro ® (2000)
27. Blueprint: The rework tax: reducing software development rework by improving requirements (2015)
28. Jarzębowicz, A., Ślesiński, W.: Survey dataset. https://www.researchgate.net/publication/324910141_RE_problems_in_Poland_-_survey_dataset

Value-Based Requirements Engineering: Challenges and Opportunities

Krzysztof Wnuk$^{(\boxtimes)}$ and Pavan Mudduluru

Department of Software Engineering, Blekinge Institute of Technology,
Karlskrona, Sweden
krw@bth.se, pamul4@student.bth.se

Abstract. In this study, we investigate the state of the literature and practice about Value-Based Requirements Engineering. We focus on identifying what models for VBRE were presented and what challenges were discussed. We triangulate our results with industrial practitioners by conducting an industrial survey with 59 respondents. We identified 26 primary and 3 secondary studies and synthesized the findings using content analysis. VBRE was identified to be having a positive impact among survey practitioners. However, challenges like aligning product, project and organization opinions, selecting a most valuable requirement for a particular release, and including time-dependent requirements were identified to be impacting the organizations. The results from the study also suggest that, value dimensions like stakeholder value and customer value were not so frequently discussed in RE processes in both literature and among our industry respondents.

Keywords: Value-based requirements engineering · Literature review
Industrial challenges

1 Introduction

Organizations facing the pressure from globalization and digitalization need to increase their competitiveness by shifting towards the value of software [1]. However, these organizations have mostly adapted software engineering (SE) in a value neutral setting where "each requirement is considered equally important, even though not all requirements are equal" [1, 2]. Researchers highlighted the need for value-based approach [3] and discussed the importance of value-based requirements [4, 5].

Requirements prioritization is the most natural moment of the RE processes where value is estimated or used to support making decisions. To overcome the challenges of prioritizing requirements, methods such as cost-value approach [6] have been developed. Wohlin and Aurum have discovered that business-oriented and management-oriented criteria and more important than technical concerns when prioritizing requirements [7]. Among other challenges that requirements engineers face when creating, measuring and managing value are: lack of metrics and selection based on release were identified in the literature [5, 8, 9].

This paper aims at classifying the value-based RE contributions, identifying the value dimensions considered in the literature and identifying the RE sub process areas

© Springer Nature Switzerland AG 2019
P. Kosiuczenko and Z. Zieliński (Eds.): KKIO 2018, AISC 830, pp. 20–33, 2019.
https://doi.org/10.1007/978-3-319-99617-2_2

where value-based RE is applied. In addition, the challenges faced by software prac-titioners in integrating value-based approach in RE are identified in a survey.

2 Background and Related Work

In the late 1990s, software development already shifted towards value-based approach. Since then value approach has emerged within software domains [1]. Figure 1 shows a model developed by Aurum and Wohlin describing how and where to capture value, including aligning business, product and project perspectives at various decisions levels to help an organization in increasing its business value [1].

Many researchers have defined value in various contexts. Khurum et al. [11] identified various value aspects from different perspectives, and gave a brief description on all the identified value constructs. The authors have created *'Software Value Map'* (SVM) which details all value aspects and sub-value aspects from four different per-spectives, they are [11]: (1) Customer perspective, (2) Financial perspective, (3) Internal business perspective and (4) Innovation and learning perspective. The following are the definitions given by other researchers: *Product Value (exchange value):* market value of the product [1], *Customer's perceived Value (use value):* how much customer is interested to pay for a product [1], *User Value:* value which users want to have [12], *System Value:* value a system has currently [12]. Over the last decade, several studies were published about VBRE [1, 2, 5, 8, 9, 12–17]. These studies have provided models, methods and frameworks that for VBRE.

Barney et al. [13] found that software product value mainly depends on the product context (i.e. where the product endures). Wohlin and Aurum [5], focused on how requirements are selected for improving the product's value. Zhang et al. [15], sug-gested to calculate value based on the customer expectations. Thakurta [14] proposed a value-based requirements prioritization approach for non-functional requirements. Lim et al. [12] proposed a value gap model which identifies the gap between the user value and system value and support requirements engineering.

Kauppinen et al. [17] proposed a set of practices that drive organizations focus towards customer value creation rather than feature development: (1) Focus on the metrics used for calculating value in RE and analyze values from these metrics, (2) Focus on other requirements activities such as value based requirements validation and value based requirements specification and (3) Focus on models that integrate VBRE.

3 Research Design and Methodology

We formulated the following research questions:

RQ1: What is the current state of the literature in VBRE?
RQ2: What metrics and models for VBRE were proposed or are used?
RQ3: What are the challenges that effect organizations in integrating VBRE?

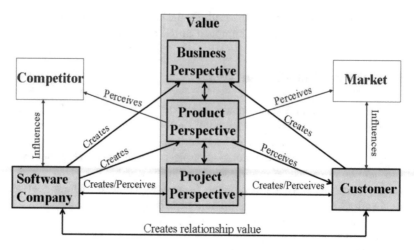

Fig. 1. Software company-value-customer triangle [1]

We used snowballing to search for relevant literature rather than a database search [18]. Database search was not considered as it heavily depends on forming a precise search string [18]. Snowballing starts with the relevant set of papers, called the start set, and analyzes references and citations in forward and backward snowballing iterations respectively. We used Engineering Village (Inspec) for identifying the initial start set papers, as suggested by Charles and Karin [19]. We selected 13 papers to the start set from 554 screened papers, S1... S13 in the reference list.

3.1   Industrial Survey

We validated and extended the SLR results in an industrial survey. The survey questionnaire was distributed electronically and designed using an online tool 'Survio. Sproull [20] identifies that the average time for the data collection is almost half when considered electronically to conventional methods. Participants for the survey were mostly requirements engineers from different organizations and had different experience levels. The survey was posted in various social networking sites such as Facebook, LinkedIn, Meetup and RE related blogs. In addition, electronic mails were sent to RE practitioners, whose email addresses were available in Software Engineering Institute (SEI), LinkedIn groups such as SARE and IREB. Additional emails were sent to authors of the articles identified from snowballing.

The survey questionnaire has a total of 15 questions divided into two parts, the first part has 4 questions and the second part has 11 questions. Demographic questions such as experience and role of the respondents form the first part of the questionnaire. The second part of questionnaire contains 11 open and close ended questions regarding VBRE. Close ended questions were included to identify the most used value dimensions and the challenges faced by the requirements engineers (synthesized from the SLR). Open ended questions were designed to identify the additional challenges, metrics and methods applied. We also investigated: (1) The impact of VBRE on an organization (positive, negative, neutral), (2) The level of agreement and impact of the challenges when integrating value approach, (3) The value dimensions mostly considered by the organizations and (4) The metrics and method used in applying VBRE[1].

The questionnaire was validated in two iterations, based on the principles provided by Kitchenham and Pfieeger [22]. The questionnaire was reviewed for issues like completeness, understandability, time taken to complete and reliability of the questions. Two subject experts and 8 software personnel were used as respondents for this pilot survey.

4 Literature Review Results

In the first backward snowballing iteration, 424 references were analyzed and 5 papers were selected for the next iteration (S14-S1Ref8, S15-S2Ref3, S16-S2Ref9, S17-S4Ref18, S18-S6Ref3). During the first forward snowballing iteration, 291 citations were analyzed, and 6 papers were included (S19-S1Cit5, S20-S3Cit14, S21-S7Cit1, S22-S7Cit4, S23-S8Cit10, S24-S12Cit5).

In the second backward snowballing iteration, 222 references were examined and no papers were added. During the second forward iteration, 270 citations were examined, and 4 papers were added (S25-S17Cit17, S26-S17Cit21, S27-S18Cit31, S28-S19Cit15). In the third backward iteration, 114 references were examined and 1 paper was selected (S29-S28Ref3). 120 citations were examined, and no paper was selected. The snowballing iterations concluded in iteration 4 with 29 papers.

29 papers were identified after four snowballing iterations. 26 papers were primary studies and three papers were secondary studies. The 26 primary studies were categorized based on the research method used in the study and type of the study, see Fig. 2. X-axis represents the research method (i.e. case study, framework, model and survey) and Y-axis represents type of the study (i.e. evaluation, solution, validation, and proposal). The classification is based on the guidelines from Runeson et al. [23] and Wierlinga et al. [24]. Out of 26 studies, 9 studies (S2, S4, S9, S11, S13, S19, S23, S24, S25) conducted case studies, 6 studies (S1, S7, S12, S18, S21, S28) proposed and evaluated various frameworks for requirement engineering activities, 9 studies (S5, S6, S8, S16, S20, S22, S26, S27, S29) proposed and evaluated models and 2 studies (S3, S17) conducted a survey to identify the important criteria for selecting the requirements from different value perspectives. A clear conclusion emerges that there are not many surveys of VBRE and justifies the need for conducting one. Moreover, most studies proposed or evaluate a model rather than a framework, suggesting rather focused approach than holistic.

The quality of the primary studies is assessed by Rigor and Relevance analysis using the guidelines of Ivarsson et al. [21]. The primary studies are divided into four categories (A, B, C, D) as per their rigor and relevance values, they are shown in Fig. 3. Out of 26 studies, 9 studies were identified as having high rigor and relevance (S2, S3,

---

[1] The survey questions are available at https://drive.google.com/open?id=1G_lmvXevWzmCGd9eIl baq-rw4nDkJMYn.

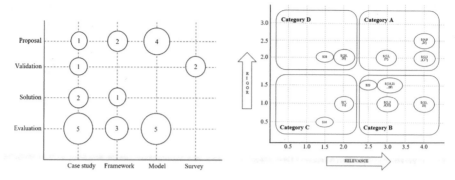

Fig. 2. Categorization of studies   Fig. 3. Rigor and relevance scores

S4, S8, S9, S12, S17, S25, S27, marked with * symbol during the thematic analysis). 10 studies were found to have high relevance and low rigor. 3 studies were identified as having low rigor and low relevance and 3 more studies have been identified as having high rigor and low relevance.

4.1 Thematic Analysis

Value-Based Requirements Engineering. Marciuska et al. [25] (S17*) explores how feature usage and customer perceived value are related. Feature usage was measured by collecting the 'nexttrailer.net' website data for two months. The results affirm that, measuring customer value using questionnaires is not very reliable as different users differently perceive value (2.5 rigor and 4 relevance).

Zhang et al. [15] (S24*), proposed a customer value model for a commercial aircraft. The model emphasized on how the initial statements from the customer are transformed into value models. The suggested process to deliver value requirements involve: (1) identifying and structuring objectives, (2) specifying attributes and constructing value models and (3) Transforming fundamental objectives into engineering characteristics (2 rigor and 3 relevance).

Babar et al. highlights the importance of stakeholder identification and prioritization in VBRE [26] (S23) and proposed a stakeholder identification framework '*Stakemeter*'. The framework was evaluated by the teams who confirmed that '*Stakemeter*' provides support in requirements engineering for developers and also for stakeholder analysis (1.5 rigor and 2.5 relevance).

Kauppinen et al. [17] (S19) highlighted the need for the shift of requirements engineering process from feature development to customer value creation. Understanding various customer groups and their perceived product value appears the be the key for VBRE. The authors have proposed 3 practices during requirements engineering process which helps the organizations in understanding customer value creation [17]: (1) Identifying potential customer groups, (2) Direct communication between practitioners and users and (3) Gathering customer processes data effectively. (1.5 rigor and 3 relevance)

Aurum and Wohlin [1] (S1) and Hasan et al. [27] (S26), discuss about the fundamental value aspects from economic theory and defines product value, customer's perceived value and relationship value from software development perspective. The study provides a '*Software Company-Value-Customer Triangle*' model which is directed towards software companies and intends to address the value.

Zhang et al. in [9] (S12) proposed an approach for qualitative and quantitative thinking about value. This approach is proposed to understand customer value from the initial statements provided by customer.

Heindl et al. [28] (S5), proposed a VBRE tool selection approach to identify the most appropriate tool support system. A feasibility study was conducted with different requirements tools at Siemens Program and Systems Engineering (PSE) to validate the approach. The results show that the approach was found to be effective in identifying the correct tool support. (1 rigor and 3 relevance).

Summary: There appears to be clear understand that value should be driving RE activities (S1, S5, S12, S19, S26). Despite that, only Zhang (S24) proposes a model that can be applied in an organization. Interestingly, stakeholder identification and prioritization (S23) and gathering customer group data (S19) are important enablers for VBRE. These extend our previous findings regarding stakeholder roles in decision making and stakeholder identification challenges in decision making [35].

Value-Based Requirements Prioritization. Racheva et al. conducted a case study on eight software organizations to identify how business value is delivered and how agile prioritization is done. Most of the participants mentioned negative value, defined as "how much value it would detract from the product's value, if the developers would not implement a feature" [29]. In addition to that, the factors such as size of the client's organization and project's size were also significant for requirement prioritization [29] (S9*). (2.5 rigor and 4 relevance)

Ramzan et al. proposed a '*Value Based Intelligent Requirement Prioritization* (VIRP)' technique. The authors have conducted an experiment on 10 projects with different parameters and constraints, concluding that it performed well in all projects, except those with unclear requirements, VIRP could not provide quality prioritization [30] (S8*). (2 rigor and 4 relevance). Sher et al. [31] (S27*), identified that the success of Value Based Software (VBS) depends on how the requirements are prioritized. It was also identified that there are more than 49 techniques described in the literature for requirement prioritization. (2 rigor and 3 relevance)

Azar et al. [8] (S18) proposed a value-oriented prioritization (VOP) framework to help in understanding the business values and maintaining an agreement between the stakeholders. (1 rigor and 4 relevance). Mohamed et al. [32] (S20), combined two prioritization techniques VOP and Hierarchical Cumulative Voting (HCV). Based on the strengths and weakness on those models, the authors proposed a new technique Value-Oriented HCV (VOHCV). Based on the findings from case study it was identified that this technique helps product managers to identify all the aspects which effect a particular requirement. (2 rigor and 2 relevance)

Thakurta [14] (S11), proposed a six-step process framework to prioritize nonfunctional requirements, this framework was validated using a case study. The results from the case study show that the framework helps to prioritize requirements based on value, which satisfy the business objectives of an organization. (1 rigor and 2 relevance)

Summary: The results suggest that some extensions to known requirements prioritization exist and take value into consideration, e.g. S18 and S20. When it comes to prioritizing quality requirements, study S8 highlights its challenging nature while study S11 offers a solution. Interestingly, negative value surfaced as an important aspect (S9) and since is highly related to reverse quality of the KANO model [36], we believe it should be further explored.

Value-Based Release Planning. Barney et al. [13] (S2*), conducted an empirical study to understand how product value is created through requirements engineering process. They also examined how practitioners recognize value of the requirements during release planning. The results show that stakeholders' prioritization and customer responsible for giving the requirement as the most important criteria for the first product, while for the second product the two most important criteria were, is the promised function delivered and the competitors' status of the requirement. The authors conclude that value creation is mostly dependent on the market in which the product exists. Equally, product value also depends on the project's maturity and what product is being developed (2 rigor and 4 relevance).

Wohlin and Aurum conducted an industrial survey to identify the most important criteria for requirement selection and how they affect value-based approach in requirements process. The results obtained from the two companies indicated that customers' perspective was important i.e. requirements issuer and priority of requirement. However, issues like delivery date and product cost-benefit were also found to be equally important. The priorities of the respondents changed when they were asked about the future scenario, the least prioritized criteria in present situation was also given equal priority in the future [5] (S3*). (2 rigor and 4 relevance)

Wohlin and Aurum discussed about which requirements are to be included in a particular release. A questionnaire was prepared to get the knowledge of various decision making criteria used in industry when a particular requirement is to be included in a release. Thirteen criteria were identified from three different perspectives (i.e. business, management and system) and sent to 8 companies to know how they prioritize their requirements and based on what criteria. The results showed that competitors, stakeholder priority of requirement, development cost-benefit and delivery date/calendar time as the most important criteria respectively. (S17*).

Value-Based Requirements Elicitation. Proynova et al. [33] (S29), proposed a three-step process for value-based elicitation by merging the existing elicitation techniques. The first step is to elicit values based on the questionnaire, the second step was to identify the attitudes of the respondents towards value and the final step is to propose requirements to related attitudes. (1.5 rigor and 3 relevance). Lim et al. proposed a value gap model which elicits the most valuable requirements from the user perspective. The model measured the value gap between users' value and system value i.e. what a system understands as user value. A case study was conducted in a company with 60 participants in Korea, using floating model the participants identified the value gap for different components. [12] (S6). (1 rigor and 3 relevance).

Summary: From the results it appears that only one paper (S29) supports requirements elicitation without having developed a system. Thus, more work is needed to suggest further support for value-based requirements elicitation.

Requirements Engineering Sub-process Areas Addressed in the Studies. The sub-process areas considered in this study are provided by Abran et al. [10] and depicted in Fig. 4. Some authors did not mention any particular RE activity or sub-process area rather they just mentioned RE, so these were marked as requirements engineering in Fig. 4. Fourteen studies [S1] [S5] [S10] [S12*] [S13] [S14] [S15] [S16] [S19] [S23] [S24] [S25*] [S26] [S28] focus on overall value rather than mention any sub-process area, e.g. business value, customer value, product value, project value and stakeholder value.

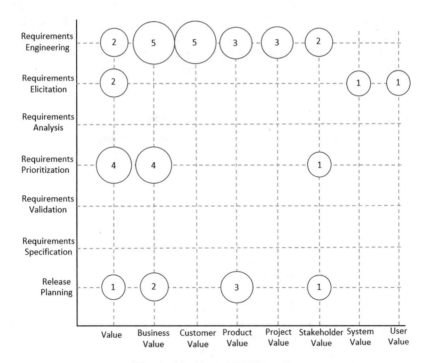

Fig. 4. Mapping of VBRE studies

The number of studies which contribute towards value-based requirements elicitation is quite low. Only 2 studies [S29] [S6] discuss this sub process-area. Nine studies were identified in value-based requirements prioritization sub-process area [S7] [S8*] [S9*] [S11] [S18] [S20] [S21] [S22] [S27*]. Five studies [S2*] [S3*] [S4*] [S17*] [S18] focused on release planning sub-process area, these 5 studies discuss value, business value, product value and stakeholder value aspects.

Summary: There were no studies found in two sub-process areas value-based requirements validation and value-based requirements specification and only one paper discussed about value-based requirements analysis. This is interesting since many authors highlight the importance of RE being data-driven, but without support for requirements validation it is unclear what data should be collected.

Based on the value dimensions described by Khurum et al. [11], we classified the studies into: business value, customer value, product value, project value, stakeholder value, system value and user value. We found that 11 studies report business value, 7 studies report customer value, 6 studies emphasize on product value, 3 studies report project value, 4 studies report stakeholder value and one study discussed both user value and system. While most of the studies discuss one or more value dimensions, 8 studies were identified which mention only value and not any one value dimension clearly, see Fig. 4. What is surprising is that not many studies discussed product value.

Challenges Identified. The challenges identified in the literature (15 studies) are analyzed and grouped based on their occurrence in RE activities and summarized in Table 1. What appears to be evident from Table 1 is a lack of challenges around requirements validation and requirements specification. According to these data, we can infer that specification should not be very challenging but validation remains an unoccupied research gap that requires more research effort and focus.

Table 1. Identified challenges and the level of occurrences in the survey (the last 4 rows)

RE activities	Challenges	ID	High	Medium	Low	Don't know
Release planning	**CH1**: Aligning product, project and organization opinions	S1, S4	29	13	11	0
	CH2: Interdependencies in requirements	S22	22	19	10	3
	CH3: Selecting a most valuable requirement for a particular release	S2, S3, S17, S13	22	16	11	4
	CH4: Revising the release plan based on the changes in requirements	S17	6	18	27	2
	CH11: Impractical expectations	S28	6	12	32	3
	CH12: Lack of criteria for value-based requirements selection	S2	7	17	29	0

(*continued*)

Table 1. (*continued*)

RE activities	Challenges	ID	High	Medium	Low	Don't know
Requirements traceability	**CH9**: Insufficient resources	S18	4	19	29	1
Requirements prioritization	**CH5**: Prioritizing the requirements based on value	S8, S11, S18, S22	10	24	19	0
	CH10: Eliciting and accommodating stakeholders' value recommendations	S3, S17	3	30	19	1
	CH13: Lack of criteria for value-based requirements prioritization	S2	3	20	28	2
Requirements elicitation	**CH16**: Including time-dependent requirements	S17	9	22	19	3
	CH7: Change in requirements	S6	6	29	18	0
Requirements analysis	**CH8**: Incomplete requirements	S6	8	14	31	0
	CH6: Determining the value of a product requirement	S1, S26	16	15	21	1
	CH14: Understanding and implementing customer value in cost-effective manner	S19, S28	5	18	30	0
	CH15: Understanding and implementing business value considering time and resources constraints	S14	11	14	28	0

5 Survey Results and Analysis

We received 68 responses, 4 respondents did not complete the survey and 5 respondents had no experience with RE. Therefore, these responses were removed, and the remaining 59 responses were included. The completion rate was 87% and as per Kitchenham et al. [34] this rate is sufficient enough to consider the results.

Demographics. Among 59 respondents, there were 15 requirements engineers (25%), 10 requirements managers (17%), 11 requirements analysts (19%), 6 project managers (10%), 5 product managers (9%), 6 developers (10%), 2 quality assurance personnel

(3%), 2 business analysts (3%) and others 2 (4%). 34% respondents were working in very large-scale organizations, 31% in large scale organizations, 17% in medium scale organizations, 14% in small scale organizations and 4% in very small scale organizations. 42% respondents were found to have more than 5 years' experience in RE, 19% respondents have 3–5 years' experience, 31% respondents have 1–3 years of experience and 8% of respondents have less than 1 year of experience. 83% (49 responses) of the respondents were identified who work in value-based context (i.e. where all the requirements are not considered equal), 10% (6 responses) of the responses were identified under value neutral setting category (i.e. where all requirements are considered equal even though they are not) and 7% (4 responses) of the respondents were not able to identify their working style.

Results. Business value was considered as the most common value dimension (mentioned 33 times), followed by customer value (21 times), stakeholder value (15 times) product value (12 times) and value (13 times). Product value was considered 12 times and project value 10 times. User and system value were considered 2 times each. Interestingly, our survey respondents presented mostly internal-business focused view on value, partly neglecting the fact that customers are the main source of profit as they pay for the product based on the perceived value. Therefore, considering both internal-business and customer perspectives on value is essential [11] as what company internally considers as value may not be value in customers' eyes.

Regarding the RE activities in which value approach is most applied the survey respondents mentioned requirements prioritization (32 times), release planning (28 times), requirements analysis (23 times), requirements elicitation (14 times) requirements specification (13 times) and requirements validation (11 times). Requirements validation is an area not identified in the SLR, leaving the RE practitioners without systematic support on how to validate if the requirements prioritized or selected based on value principles are appreciated by the customer or users.

Table 1 shows the level of occurrence of each challenge from the responses (the last four rows). The most frequently occurring challenges are CH1, CH2 CH3. This supports the SLR results and highlights that industry still needs support in prioritizing requirements and planning releases based on value. We also believe that only after sufficient improvements in release planning and prioritization, other areas can be also improved with value-based approaches.

Statistical Tests: The results from the Chi-square test showed no relationship between the experience of the survey participants and any of the challenges. No relation was also identified between the respondents' role and the level of occurrence of the challenges. As there was no relationship identified from the descriptive analysis, non-parametric test was used to find the mean rank and standard deviation of the occurrences (Friedman test). From the non-parametric test, challenges like aligning product, project and organization opinions (CH1), selecting the most valuable requirement for a particular release (CH3) and including time-dependent requirements (requirements with time as an attribute and requirements that only have value at a certain time) (CH16) were identified as the most effecting. Challenges like impractical expectations (CH11) and lack of criteria for value-based requirements prioritization (CH12) were identified as the least occurring.

Finally, 74% (44 responses) of respondents identified that VBRE has a positive impact on organizations, 10% (6 responses) of respondents identified that VBRE has no impact on the organizations, only 2 respondents felt that VBRE has a negative impact on the organization and 12% (7 responses) of respondents were unable to identify the impact of VBRE on organizations.

Summary: The results of the industrial survey not only confirm most of the findings from the SLR but also emphasize that introducing value-based RE is a significant improvement activity that should be gradually performed. Companies still struggle to introduce value into requirements prioritization and release planning and these should be the first improvement areas.

6 Conclusions

This work explores the state of the literature in VBRE and conducted an industrial survey to triangulate the findings from the literature with practitioners' viewpoints. We focused on methods or models for VBRE, challenges and phases of RE processes that have the need or are currently supported by value-based thinking. We followed a systematic and rigid process for literature search and piloted our survey instrument before data collection.

Answering RQ1, we examined the literature in VBRE and conclude that case studies and models dominate the papers, mostly with high relevance or high or low rigor. 14 studied discussed general value in RE, 9 studies focused on value-based requirements prioritization and 5 studies in release planning. The literature review also identified 15 challenges of VBRE.

Regarding RQ2, requirements prioritization and release planning were the most frequently mentioned RE phases among our survey participants. Interestingly, the survey results suggest that business value is more important than customer value, neglecting the integral nature of the two concepts and the driving force that the customers have in delivering profitability to an organization.

Regarding RQ3, the challenges found in the literature were evaluated by the survey respondents. The respondents prioritized aligning product, project and organization opinions, interdependencies in requirements and selecting a most valuable requirement for a particular release.

In future work, we plan to create a framework for holistically considering value in requirements engineering sub-processes and areas. We believe that it is critical to introduce value-based thinking as early as stakeholder or user group identification to filter valuable requirements sources from other requirements sources.

Acknowledgements. This work is supported by the EASE Industrial Excellence Center Phase III project founded by VINNOVA as well as Sony Mobile Communications, Axis Communications and Softhouse Consulting.

References

1. Aurum, A., Wohlin, C.: A value-based approach in requirements engineering: explaining some of the fundamental concepts. In: Requirements Engineering: Foundation for Software Quality, pp. 109–115 (2007). **S1**
2. Barney, S., Aurum, A., Wohlin, C.: A product management challenge: creating software product value through requirements selection. J. Syst. Architect. **54**(6), 576–593 (2008). **S4**
3. Boehm, B.: Value-based software engineering: reinventing. ACM Softw. Eng. Notes **28**(2), 2–3 (2003). **S15-S2Ref8**
4. Favare, J.: Managing requirements for business value. IEEE Softw. **19**(2), 15–17 (2002). **S14-S1Ref8**
5. Wohlin, C., Aurum, A.: Criteria for selecting software requirements to create product value: an industrial empirical study. In: Value-Based Software Engineering, pp. 179–200 (2006). **S3**
6. Karlsson, J., Ryan, K.: A cost-value approach for prioritizing requirements. IEEE Softw. **14**(5), 67–74 (1997)
7. Wohlin, C., Aurum, A.: What is important when deciding to include a software requirement into a project or release. Int. Symp. Empir. Softw. Eng. **186**, 20–28 (2005)
8. Azar, J., Smith, R.K., Cordes, D.: Value-oriented requirements prioritization in a small development organization. IEEE Softw. **24**(1), 32–37 (2007). **S18-S6Ref3**
9. Zhang, C.B.X., Auriol, G., Shukla, V.: How to think about customer value in requirements engineering **24**(1), 483–486 (2011). **S12**
10. Abran, A., Bourque, P., Dupuis, R., Moore, J.: Guide to the Software Engineering Body of Knowledge-SWEBOK. IEEE Press, Piscataway (2001)
11. Khurum, M., Gorschek, T., Wilson, M.: The software value map—an exhaustive collection of value aspects for the development of software intensive products. J. Softw. Evol. Process **25**(7), 711–741 (2013). **S13**
12. Lim, S., Lee, T., Kim, S.: The value gap model: value-based requirements elicitation. In: IEEE International Conference on Computer and Information Technology, pp. 885–890 (2007). **S6**
13. Barney, S., Aurum, A., Wohlin, C.: Quest for a silver bullet: creating software product value through requirements selection. In: Software Engineering and Advanced Applications, pp. 274–281 (2006). **S2**
14. Thakurta, R.: A value-based approach to prioritise non-functional requirements during software project development. Int. J. Bus. Inf. Syst. **12**(4), 363–382 (2013). **S11**
15. Zhang, X., Auriol, G., Eres, C., Baron, C.: A prescriptive approach to qualify and quantify customer value for value-based requirements engineering. Int. J. Comput. Integr. Manuf. **26**(4), 327–345 (2013). **S24-S12Cit5**
16. Gordijn, J., Akkermans, J.: Value-based requirements engineering: exploring innovative e-commerce ideas. Requirements Eng. **8**(2), 114–134 (2003). **S16-S2Ref9**
17. Kauppinen, M., Savolainen, J., Lehtola, L., Komssi, M., Töhönen, M., Davis, A.: From feature development to customer value creation. In: 17th Requirements Engineering Conference, pp. 275–280 (2009). **S19-S1Cit5**
18. Wohlin, C.: Guidelines for snowballing in systematic literature studies and a replication in software engineering. In: 18th International Conference on Evaluation and Assessment in Software Engineering, pp. 38–46 (2014)
19. Charles, W., Karin, K.: Engineering Communication. Cengage Learning Inc., Boston (2015)
20. Sproull, S.: Using electronic mail for data collection in organizational research. Acad. Manag. J. **29**(1), 159–169 (1986)

21. Ivarsson, M., Gorschek, T.: A method for evaluating rigor and industrial relevance of technology evaluations. Empir. Softw. Eng. **16**(3), 365–395 (2011)
22. Kitchenham, B., Pfleeger, S.L.: Principles of survey research part 4: questionnaire evaluation. ACM SIGSOFT Softw. Eng. Notes **27**(3), 20–23 (2002)
23. Runeson, P., Host, M., Rainer, A., Regnell, B.: Case Study Research in Software Engineering: Guidelines and Examples. Wiley, Hoboken (2012)
24. Wieringa, R., Maiden, N., Mead, N., Rolland, C.: Requirements engineering paper classification and evaluation criteria: a proposal and a discussion. Requirements Eng. **11**(1), 102–107 (2006)
25. Marciuska, S., Gencel, C., Abrahamsson, P.: Exploring how feature usage relates to customer perceived value: a case study in a startup company. In: ICSOB 2013 Conference, pp. 166–177. Springer (2013). **S17Cit17**
26. Babar, M.I., Ghazali, M., Jawawi, D.N., Zaheer, Stakemeter K.B.: Value-based stakeholder identification and quantification framework for value-based software systems. PLoS ONE **10**(3), e0121344 (2015). **S23-S8Cit10**
27. Hasan, S.M.N., Hasan, M.S., Mahmood, A., Alam, M.J.: A model for value based requirement engineering. Int. J. Comput. Sci. Netw. Secur. **10**(12), 171–177 (2010). **S26-S17Cit21**
28. Heindl, M., Reinisch, F., Biffl, S., Egyed, A.: Value-based selection of requirements engineering tool support. In: 32nd EUROMICRO Conference, pp. 266–273. IEEE (2006). **S5**
29. Racheva, Z., Daneva, M., Sikkel, K., Herrmann, A., Wieringa, R.: Do we know enough about requirements prioritization in agile projects. In: 18th IEEE International RE Conference, pp. 147–156. IEEE (2010). **S9**
30. Ramzan, M., Jaffar, A., Shahid, A.: Value based intelligent requirement prioritization (VIRP): expert driven fuzzy logic based prioritization technique. Innov. Comput. Inf. Control, **7**(3), (2011). **S8**
31. Sher, F., Jawawi, N.D., Mohamad, R., Babar, M.I.: Multi-aspects based requirements priortization technique for value-based software developments. In: International Conference on Emerging Technologies (ICET), pp. 1–6. IEEE (2014). **S27-S18Cit31**
32. Mohamed, S.I., ElMaddah, I., Wahba, A.M.: Towards value-based requirements prioritization for software product management. Int. J. Softw. Eng. **1**(2), 35–48 (2008). **S21-S7Cit1**
33. Proynova, R., Paech, B., Wicht, A., Wetter, T.: Use of personal values in requirements engineering—a research preview. In: REFSQ 2010, pp. 17–22 (2010). **S29-S28Ref3**
34. Kitchenham, B., Pfleeger, S.L.: Principles of survey research. Softw. Eng. Notes **27**(5), 17–20 (2002)
35. Wnuk, K.: Involving relevant stakeholders into the decision process about architecturally significant assets: challenges and opportunities. In: International Conference on Software Architecture Workshops, pp. 129–132 (2017)
36. Kano, N., Nobuhiku, S., Fumio, T., Shinichi, T.: Attractive quality and must-be quality. J. Jpn. Soc. Qual. Control **14**(2), 39–48 (1984). (in Japanese)

Applying Use Case Logic Patterns in Practice: Lessons Learnt

Albert Ambroziewicz and Michał Śmiałek[(✉)] [iD]

Faculty of Electrical Engineering, Warsaw University of Technology,
pl. Politechniki 1, Warsaw, Poland
smialek@iem.pw.edu.pl

Abstract. Use cases are popular means to specify functional requirements in terms of application logic. A typical way to represent this logic
is through scenarios expressed using textual or graphical notations. Use
case scenarios can be generalised to offer abstract use cases – use case
logic patterns. Such patterns capture recurring logic of user-system interactions independent of the particular problem domain. They are composed of abstract use case models and use case scenarios formulated in
constrained natural language. Scenarios refer uniformly to an abstract
domain model. Use case patterns can be easily instantiated through substituting the abstract domain with a concrete one. In this paper we
present a study on application of use case patterns in a real industry
project. At first, the project attempted at specifying reoccurring functionality using an ad-hoc approach. This resulted in poor quality specifications as judged by experienced analysts. Following this, we have proposed to use the pattern approach which could be compared with the
previous attempt. We have used a library of patterns that were applied
through instantiating them in a particular problem domain. Lessons
learnt from this comparison show improvement in clarity, repeatability
and correctness, regardless of a tooling environment used.

Keywords: Use cases · Use case patterns · Industrial practice
Application logic

1 Introduction

Use cases, as introduced originally by Jacobson [11] constitute means to specify
goal-driven sequences of user-system interactions that express the logic of the
given application. It is a natural desire to be able to extract abstract use case
logic and reuse it in different domain contexts. Probably the first attempt at
this is the proposition by Cockburn [6] to introduce parameterised use cases.
Cockburn gives an example of the "Find a whatever" use case but leaves elabo-
ration of the idea to the readers. Later, Issa et al. [10] report high reuse potential
for use cases. Their research shows that the majority of use cases for a typical
system can be reused either within a given problem domain or independent of
the domain.

© Springer Nature Switzerland AG 2019
P. Kosiuczenko and Z. Zieliński (Eds.): KKIO 2018, AISC 830, pp. 34–49, 2019.
https://doi.org/10.1007/978-3-319-99617-2_3

Use case patterns can be expressed at the level of templates and styles for structuring use case models and writing their scenarios. Many such approaches have been advocated in literature [2,6,13]. Some propose to capture patterns through analysing repeatable arrangements of use cases. This results in specifying recurring layouts for use case diagrams, i.e. at the very highest level of detail [19]. Others propose to define patterns at the level of use case content, or in other words – "contents of the oval" (see an interesting approach by Langlands [14]). Initially, we would need to define some standardised language to specify use cases, and the language itself would be composed of certain sentence "patters" [7]. An example of such a comprehensive language that includes a constrained natural language syntax to express use cases in a standardised way is the Requirements Specification Language (RSL) [12,24]. Also, other approaches propose formalised approaches to define reoccurring logic within use cases [1,9,16]. In this context, we should also mention approaches to formalise semantics of use case scenarios. Unfortunately, semantics of use cases have raised serious criticism [21] and still raises issues [4,8]. In order to formulate patterns for use cases we need to make use cases semantically precise. In this paper, we use control flow semantics for use case scenarios resulting from our previous work [22,24]. This allows us to assume certain meaning of the pattern constructs that is universally understood by the pattern users. It also allows us to extend the approach with code generation capabilities as mentioned in the Conclusion section.

Research on use case patterns is quite sparse and most of the approaches offer only limited validation in terms of a simple case study or example. An interesting exception is the study by Ochodek et al. [18] who performed a survey-based study on usefulness of patterns in use case specification. This study shows certain threats to validity that are associated e.g. with the tools used. In the current paper we report on practical application of use case logic patterns where tool usage is an important supporting element. Our aim was to acknowledge possible levels of reuse when such patterns are applied and possible improvement in quality. Due to certain commercial restrictions we could collect only anecdotal results. Still, these results can contribute to understanding the role of patterns in early phases of software development.

In the following sections we first present an introduction to the language and pattern library used in the study. Next, we present a referential approach initially used to develop use case models. Following this, we show an alternative method when use case logic patterns are applied. Finally, we conclude by discussing the two approaches and comparing them.

2   Use Case Logic Language

In the study we use RSL as the use case logic language. RSL is to some extent based on UML [17] but introduces significant improvements regarding detailed syntax for use case scenarios and general semantics of use cases. Figure 1 shows an example RSL specification. At the top, we need to define a use case model, similar to how it is done using UML. The main difference in RSL is that the «include»

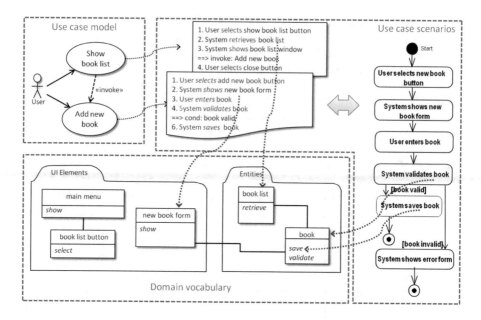

Fig. 1. Example model in RSL use case logic language

and «extend» relationships are substituted by the single «invoke» relationship. Unlike for the UML counterparts (see e.g. work by Simons [21]), «invoke» has precisely defined function-call semantics [22].

Use case semantics is also reflected in the syntax for use case scenarios. (see ceter-top in Fig. 1). A typical sentence in a scenario defines a single action performed either by an actor or by the specified system. Some sentences are control sentences and define certain conditions and invocations. The invocation sentences are similar to function calls where another use case logic is called. Scenarios can be also expressed in graphical form as activity diagrams. This form is especially suitable to show various alternative flows through a use case.

The key element in an RSL specification is its strict reliance on a central domain vocabulary (see bottom of Fig. 1). Every action sentence in a scenario consists of a subject (noun), a verb and one or two objects (nouns). The nouns are represented by domain notions expressed similarly to UML classes. The verbs constitute phrases (operations) within these notions. When looking at Fig. 1 we can notice that by substituting the notions in the vocabulary we can easily change to another problem domain. For instance, the "book" notion can be change to the "shop item" or to a "whatever" as postulated by Cockburn [6]. This gives us the opportunity to formulate abstract use case logic independent of any specific problem domain. We only need to substitute a concrete domain vocabulary with an abstract one.

The RSL allows to build precise specifications but in order to propose use case logic patterns we need to extend it. The extensions are based on our previous

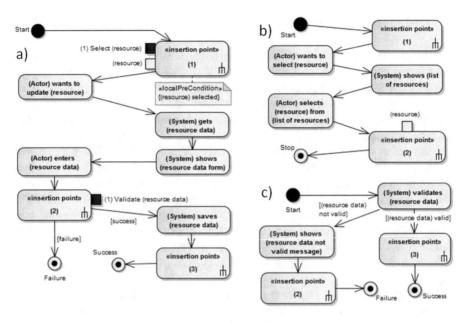

Fig. 2. Example pattern "snippets": (a) Update (resource), (b) Select (resource), (c) Validate (resource)

work [3], and an example illustration is given in Fig. 2. The idea is to be able to define short "snippets" of use case logic, of which one can build larger patterns. The figure presents two of such snippet patterns related to each other. The "Update (resource)" pattern start its logic by inserting the logic of the "Select (resource)" pattern (see black pin). The two patterns are bound through the abstract "(resource)" notion (see white pin). The "Validate (resource)" pattern is inserted in a similar way. Altogether, this results in a complete abstract logic for updating data of some abstract resource (a book, a shop item etc.). As we can see, the notation is a variation on activity notation present in UML. The pins serve as parameters to "insertion points" which allow to combine various snippets. Numbers associated with insertion points allow for unambiguous definition of parameter links between patterns.

Following such notation and approach, we can build a library of patterns. For instance, we can define a CRUD (Create-Read-Update-Delete) pattern, where the above "Update (resource)" in one of the elements. Other such simple patterns can include resource transfer, resource sharing, resource partition, binding of resources, searching for resources, managing processes and so on. In fact, in our previous work [3] we have proposed a ready set of such patterns. In the following sections we will present and evaluate an approach to extend the library and apply it to a specific system that exhibits reoccurring application logic.

3   Reference Approach

To validate the use case pattern approach we have used a significantly sized real-life project. The subject of this study project was a significant, governmental IT system[1] to which we had access and could observe and consult the requirements team. The system was to consists of several subsystems, including typical security and integration components, but mostly domain-dependent data interchange subsystems, a portal platform, a specialized data warehouse and a registers subsystem. During the study we concentrate on the registers subsystem which was particularly suitable as the subject of pattern application. This subsystem is an important, core part of the overall system. It holds several registers that contain data records regarding people, items, documents and events. This allows to centralise the problem domain elements and unify relations between other subsystems. Data in the registers can be easily verified for correctness and also facilitate interoperability with external systems, at the same time ensuring high quality – mainly coherence and validity – of other data processed by the system.

Functionality of the registers subsystem is quite complex, as the registers have varied characteristics. There are four main registers in the system and several smaller ones. The requirements team thus concluded that the best way to approach the problem was to create a repeatable "template" use case model that could be specialised for each of the registers. Thus, the requirements team assumed the following modelling principles. A use case model should be created for each subsystem, even when a subsystem does not offer human interfaces and interacts only with other systems via automated interfaces. Such a use case model according to the modelling principles would follow most of the core RSL concepts: application (use case) logic and its domain vocabulary are separated but tightly linked, use case scenarios are described unambiguously using SVO sentences, scenarios are expressed textually and diagrammatically. However, the team decided to use standard UML relationships of «include» and «extend», instead of the RSL-specific «invoke». This was due to that the UML relationships were assessed as more standard and familiar to the requirements team members.

The use case model was defined using a standard UML CASE tool (Enterprise Architect) but equipped with an elaborated scenario editor. This allowed the team to develop structured scenarios and generate compliant activity diagrams automatically within the tool. In addition, the developed use case scenarios would contain explicit exception handling. This meant that some of the alternative scenarios were intended only to provide descriptions of exceptions handled at a technical level (like failed data reads), not directly resulting from business decisions or conditions. Finally, the team has assumed certain rules for structuring the domain vocabulary model. This model contained notions used in scenarios but also served as a global glossary for the whole project. It generally

---

[1] Permission to use the name of the system or the organization was not given due to commercial issues. As a result, all the examples were generalized and detached from the original business domain of the study.

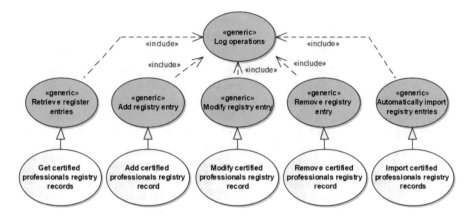

Fig. 3. Illustration of the initial approach: registry management

included just the noun-related elements, omitting the verbs. In this respect, the model was not fully conformant to RSL.

Considering the above general rules, the requirements team has also made decisions regarding ways to shape the use case model. This was done to assure uniformity of the model developed by several analysts. The approach taken, based on standard UML constructs, was to create a generic use case model, from which register-specific use case models were derived using UML's specializations. This is illustrated in Figs. 3 and 4, where the upper (green colour) use cases are the generic ones. The figures show use cases that allow for generic management of a registry. This includes CRUD operations (here called: Add-Get-Modify-Remove) on registry entries, listing of entries (Retrieve) and automated import from an outside (sub)system.

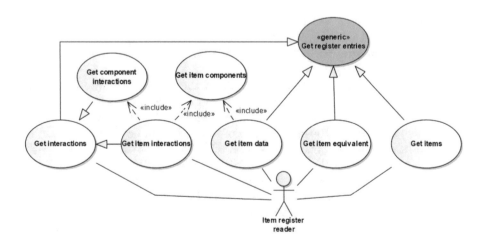

Fig. 4. Illustration of the initial approach: registry access

For each specific registry, a model was created that derived from the above generic model. One of the requirements analysts took the generic model and adjusted it to the specific requirements. This consisted in creating generalisation relationships (like in Figs. 3 and 4), changing use case names and their scenarios. Changes to scenarios involved to large extent manual substitution of the notions used in the generic use case model, like "registry" or "entry", with specific ones like "certified professionals registry" or "certified professionals record". This activity also included modelling of the system's actors. At this stage, also additional (non-standard) use cases were introduced. This was done from scratch or using the general use cases as a starting point. The resulting specification resulted in approximately 150 use cases, developed in 3 teams consisting of 3–4 requirements analysts each. It should be noted that such a practice was quite common for the team as it followed typical approaches used in the contemporary software industry. The team was not acquainted with any formalised pattern-oriented use case approach. This can be seen as a typical situation occurring during the requirements phase, illustrating the fact that requirements (use case) patterns have never really gained momentum in the industry.

The resulting models were subject to quality review which revealed serious problems. The main source of errors was different understanding of the generalization relationship by different analysts, as its semantics is not specified with necessary precision in UML [15]. Some of the analysts just copied the user-system interaction descriptions from the general use cases. In some cases this involved changing general notions (like "register") to specific ones (like "people index"). This generated many logical errors, due to that the analysts where too focused on copying and updating text rather than applying it in the new context. Also, this copy-paste approach caused various inconsistencies as not all abstract notions were replaced with their specific counterparts. There were some analysts (a minority), that did not copy scenario sentences, but rewrote them referencing the general ones like in: "steps 1–4 as in the generalization, step 5. System/Actor does something". This approach was less time-consuming for the analysts and produced more concise results. However, it has still caused the same language inconsistencies and logical errors as the previous one. Namely, there was no substitution of the general notions with the specific ones, and some of the specific requirements were forgotten.

Many of the analysts have copied parts of the general use case behaviour, added some sentences after it, copied/referenced some more sentences, afterwards inserted some actions here and there. Such a technique was considered error prone, as it breaks the original flow of interaction in ways not expected by the person modelling the general interaction. In practice, it turned out that – for example – some validity checks in the registers functionality were not described, unnecessary multiplied or done as originally planned, but without triggering alternative paths in the scenarios. What should be emphasized is that the above described mistakes were not always apparent, especially in the cases where notion names were not substituted properly. For example, such a situation was discovered in a use case that defines interaction between many registers. Improper substitution caused mismatches in the individual register names, significantly confusing the scenario readers.

The quality review has also revealed other issues not mentioned here for brevity. All these mistakes can be perceived as "technical" and possible to avoid by careful analysts. Still, on the other hand, it seems that they are so common and obvious that they should be expected given the context and the human nature. If a person copies large amount of uniform, tedious text, as in use case scenarios, mistakes are very common. Several works document this fact: whether it is programming code [5] or any other structured document [20]. Mechanical text replacing (like notion name substitution) makes the mistake of replacing wrong text highly possible. Also, copying text over and over limits ones critical attitude towards the copied message. It becomes perceived as being one's own text while it was in fact written by someone else. Moreover the person gets used to even very visible mistakes and copies them unconsciously.

4   Use Case Pattern Approach

To solve the identified problems, we have proposed to use the use case pattern approach. This included defining the "Manage (register)" pattern presented in Figs. 5 and 6. As it can be noted, the pattern follows strictly the RSL notation and uses RSL's pattern-oriented extension as presented in the previous sections. This pattern was to be reused as a starting model for every register in the system.

As it can be noted, the "Manage (register)" pattern is based on some of the elements in the original use case pattern library. The abstract logic of the pattern consists of 5 use cases and 2 actors (one general, one specific). The first

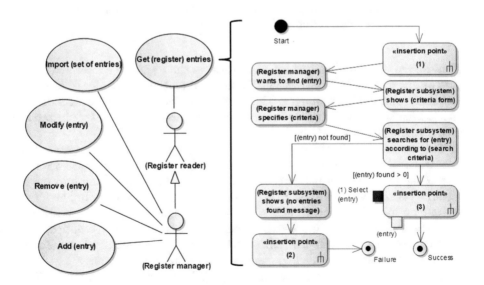

Fig. 5. The "Manage (register)" pattern: use cases and their contents

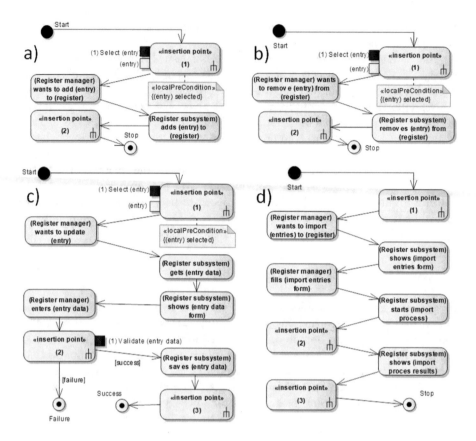

Fig. 6. Contents of the "Manage (register)" pattern: (a) Add (entry), (b) Remove (entry), (c) Update (entry), Import (set of entries)

of the actors is the register reader, who retrieves the register's entries according to some given criteria. This functionality is based on the "Search for (resource)" pattern from the original library. The abstract notion "resource" is substituted with the concrete "entry". This in turn is an abstract notion for other instances of the "Manage (resource)" pattern.

The second actor in the model is responsible for managing the register. The actor can evoke use cases that allow to add, remove and delete entries in the register. They are based on patterns from the original library grouped into larger patterns called "Manage (resource)" and "Manage (collection)". In this case, abstract notions of "resource" and "collection" are replaced by "entry" and "register" respectively. Figure 6 presents scenarios of this part. We can notice, that the scenarios refer to "snippet" patterns from the original library but mapped onto new vocabulary notions (e.g. "Select entry" vs. "Select resource").

To create the new pattern for the study we have followed the process illustrated in Fig. 7. The goal is to create the "Manage (register)" pattern as an

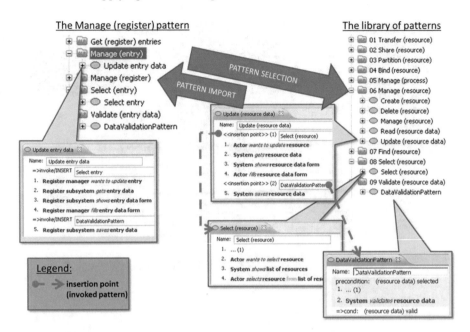

Fig. 7. Creation of the "Manage (register)" pattern in the ReDSeeDS tool

intermediary for other – specific registry models. In the presented example, an analyst intends to use a suitable pattern from the original pattern library (see the "Pattern selection" arrow pointing at the "Manage (resource)" pattern that includes the "Update (resource data)" use case snippet. The pattern is imported (see the "Pattern Import" arrow) using a dedicated tool: in this case it is the ReDSeeDS tool (www.redseeds.eu) [25]. The dashed arrows in the figure indicate patterns inserted into the chain of imported logic. For instance, "Manage (resource)/Update (resource data)" inserts "Select (resource)" and "Validate (resource data)". Such pattern import operations are repeated until the base "Manage (register)" pattern is completely formed (following the pattern relationships to the library patterns as described above). The next activity at this stage consisted of writing additional use cases (such as "Import (set of entries)") and adjusting the ones imported from the library.

The second step of the pattern adaptation process is presented in Fig. 8. At this stage, the use case is first selected from the "Manage (register)" pattern and instantiated for the concrete register (see the "Pattern import" arrow). Please note the reused scenario sentences containing all the abstract notions substituted with concrete ones (abstract scenario connected by green arrow with a concrete one in the Figure).

Along with instantiating the abstract use case scenarios, also the abstract domains are aligned. It can be noted, that in this example, the domain adaptation is performed similarly to logic adaptation, in two steps. First, the parts of the original library's abstract domain associated with the patterns used in this study

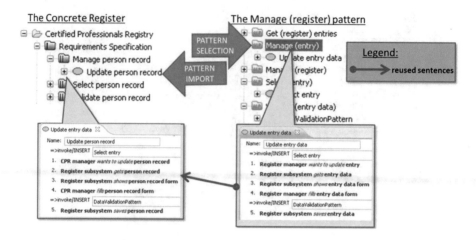

Fig. 8. Instantiation of the "Manage (register)" pattern

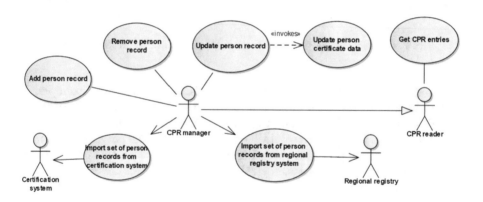

Fig. 9. Use cases for CPR management as instances of "Manage (register)"

are aligned to comply with the parts of the abstract domain associated with the "Manage (register)" pattern. Finally, the resulting domain is instantiated to become the concrete register domain.

We illustrate the resulting pattern instances for the Certified Professionals Registry (CPR). It is a secondary-type register, which groups data about professionals in a given domain. Its use case model is presented in Fig. 9. The model is a quite straightforward adaptation of the Register management pattern in regard to adding, deleting and retrieving entries. The pattern logic was supplemented with the functionality of modifying register entries (in the professionals registry, the person's certificate is modified independently of personal information data about a given professional). The main functional area where modifications were

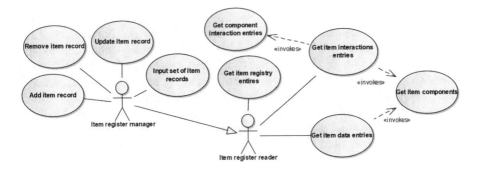

Fig. 10. Use cases for item management as instances of "Manage (register)"

applied to the original pattern logic were use cases defining entry import capabilities of the registry. These functional units realized the requirements stating
multiple and different primary and secondary data sources for the registry (modelled by separated use cases and secondary actors for each data supplier).

A second example instance of the "Manage (register)" pattern is the Items
Registry. Items are compound elements that can have components and several
interaction entries. The use case model (slightly simplified) for the items registry is presented in Fig. 10. As previously, the functionality for the register is a
direct adaptation of the base "Manage (register)" pattern – the use cases allow
adding, removing, updating register records. The major change in the instantiated model is in the area of retrieving the items stored in the register. Items can
be retrieved from the register through the search criteria, exactly as in the base
pattern. However, there are also several other instances of the "Get (register)
entries" snippet. This involves retrieving item components and interactions. This
illustrates flexibility of the pattern approach to define similar application logic
for different data.

5   Results and Discussion

The outcomes of applying the presented pattern mechanism indicate resolv-
ing most important problems caused by the original (reference) approach. The
resulting models were consistent in regards to the wording used: all the notions
names were substituted properly. The same names for important business entities were used consistently throughout the whole model (not only locally – by a
use case model relevant to a given register, but in all model elements/packages).
The number of logical mistakes caused by use case interleaving and copy-paste
operations was minimized, as changes done to the original interaction sequences
could be made only in the places marked by the pattern "insertion points" and
guarded by precisely defined local pre-conditions. Importantly, the copy-paste
mechanism was substituted by domain substitution controlled by a dedicated
tool. This way, all the notions were consistently applied in the new domain context.

Application of use case patterns resulted in noticeable productivity gains. The analysts involved in the study have indicated that the pattern instantiation process is much more "manageable" than the previous approach to specialise use cases. One of the aspects reported was associated with the registry names that were quite long and complex (consisting usually of several words). Their manual substitution necessitated significant effort (apart from possible mistakes) that was almost completely removed during tool-supported pattern instantiation. The analyst had to only define the new domain notions that were automatically substituted in the instantiated scenarios. What is also important, the substitution was not done through a simple text search but involved keeping the sentence parts linked with the domain vocabulary. This also allowed for multi-step instantiation as presented in the previous paragraphs.

In the study we have also compared using two different tools: a standard UML tool (Enterprise Architect) and a dedicated tool (ReDSeeDS). Both tools were equipped with mechanisms facilitating pattern instantiation. However, the ReDSeeDS tool has built-in mechanisms that allow for instantaneous substitution of scenario sentence parts when the associated domain vocabulary changes. Moreover, it is equipped with a dedicated wizard, as illustrated in Fig. 11b. In the case of Enterprise Architect, the instantiation mechanism uses an extended paste option. In this case, the tool offers a dialogue that allows the analyst to modify sentence text individually during the paste operation, as illustrated in Fig. 11a. In both cases, the most time consuming element in pattern instantiation was substitution of the problem domain. In both cases it was perceived as significantly faster than a typical copy-paste approach. It is so because all the elements are copied and pasted at once, as opposed to typical text operations, where the copied text has to be moved individually, section by section. This advantage is less evident for the approach with the standard UML tool, where much of the domain instantiation had to be repeated for all the use cases.

The results stemming from our study are encouraging also regarding the quality of the output model. This was assessed during quality reviews as significantly better than in the reference approach. It can be also noticed that quality improvement was not achieved at the expense of increased workload on the analyst side. On contrary, their effort decreased as indicated above. Moreover, the

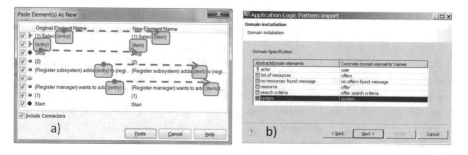

Fig. 11. Pattern instantiation using tools: (a) Enterprise Architect, (b) ReDSeeDS

study shows that the use case pattern approach can be applied in a modelling environment not fully supporting pattern instantiation. This is especially important in typical industry settings where a specific tooling environment is enforced by e.g. company policies. In this case, standard CASE tool mechanisms can be applied to obtain functionality close to the desired automated pattern instantiation. A typical CASE tool with "smart" copy/paste capabilities can be sufficient for this purpose.

6 Conclusion and Future Work

The presented study gives initial insights on applicability of use case patterns in industry practice. It shows that consistent usage of patterns in specifying use case contents can reduce effort and increase quality of requirements. The study is definitely not conclusive regarding levels of improvement in these two aspects. Currently, we offer anecdotal evidence based on discussions with requirements analysts involved in the study. More detailed and systematic results would necessitate setting-up a full-sized comparative study, and significant analyst involvement and investment. However, this is very difficult in industry settings. We treat this as future work which would also involve improving the pattern instantiation mechanisms.

We can note that use case patterns expressed using the presented approach have significant potential for automatic processing. In our previous work, we have researched translation of RSL models (including scenarios) to operational code. We have defined precise runtime semantics of scenarios [22]. This resulted in automatic generation of application prototypes directly from precisely formulated requirements specifications [23,25]. Future work would thus involve extending this approach also to use case patterns. Possible directions for future research would involve creating a searchable and extendable library of patterns with integrated application generation capabilities. The developers would be able to quickly generate application prototypes from readily defined application logic specifications.

References

1. Aballay, L., Introini, S.C., Lund, M.I., Ormeno, E.: UCEFlow: a syntax proposed to structuring the event flow of use cases. In: 8th IEEE Computing Colombian Conference, pp. 1–6 (2013). https://doi.org/10.1109/ColombianCC.2013.6637517
2. Adolph, S., Bramble, P., Cockburn, A., Pols, A.: Patterns for Effective Use Cases. Addison Wesley, Reading (2002)
3. Ambroziewicz, A., Smialek, M.: Application logic patterns - reusable elements of user-system interaction. In: MODELS 2010. Lecture Notes in Computer Science, vol. 6394, pp. 241–255 (2010). https://doi.org/10.1007/978-3-642-16145-2_17
4. Astudillo, H., Génova, G., Śmiałek, M., Llorens Morillo, J., Metz, P., Prieto-Diáz, R.: Use cases in model-driven software engineering. Lecture Notes in Computer Science, vol. 3844, pp. 262–271 (2006). https://doi.org/10.1007/11663430_28

5. Chou, A., Yang, J., Chelf, B., Hallem, S., Engler, D.R.: An empirical study of operating system errors. In: Symposium on Operating Systems Principles (2001). https://doi.org/10.21236/ada419594
6. Cockburn, A.: Writing Effective Use Cases. Addison-Wesley, Reading (2000)
7. Díaz, I., Losavio, F., Matteo, A., Pastor, O.: A specification pattern for use cases. Inf. Manag. **41**(8), 961–975 (2004). https://doi.org/10.1016/j.im.2003.10.003
8. Génova, G., Llorens, J., Metz, P., Prieto-Díaz, R., Astudillo, H.: Open issues in industrial use case modeling. In: Jardim Nunes, N., Selic, B., Rodrigues da Silva, A., Toval Alvarez, A. (eds.) UML Modeling Languages and Applications, pp. 52–61. Springer, Heidelberg (2005). https://doi.org/10.1007/978-3-540-31797-5_6
9. Georgiades, M.G., Andreou, A.S.: Patterns for use case context and content. In: Proceedings of the 13th International Conference on Software Reuse, pp. 267–282 (2013). https://doi.org/10.1007/978-3-642-38977-1_18
10. Issa, A., Odeh, M., Coward, D.: Using use case patterns to estimate reusability in software systems. Inf. Softw. Technol. **48**, 836–845 (2006). https://doi.org/10.1016/j.infsof.2005.10.005
11. Jacobson, I., Christerson, M., Jonsson, P., Övergaard, G.: Object-Oriented Software Engineering: A Use Case Driven Approach. Addison-Wesley, Reading (1992)
12. Kaindl, H., Smialek, M., Wagner, P., et al.: Requirements specification language definition. Project Deliverable D2.4.2, ReDSeeDS Project (2009). www.redseeds.eu
13. Kulak, D., Guiney, E.: Use Cases: Requirements in Context, 2nd edn. Addison Wesley, Reading (2012)
14. Langlands, M.: Inside the oval: use case content patterns. Technical report, Planet Project, v. 2 (2014)
15. Metz, P., O'Brien, J., Weber, W.: Against use case interleaving. In: UML 2001. Lecture Notes in Computer Science, vol. 2185, pp. 472–486 (2001). https://doi.org/10.1007/3-540-45441-1_34
16. Misbhauddin, M., Alshayeb, M.: Extending the UML use case metamodel with behavioral information to facilitate model analysis and interchange. Softw. Syst. Model. **14**(2), 813–838 (2015). https://doi.org/10.1007/s10270-013-0333-9
17. Object Management Group: OMG Unified Modeling Language, version 2.5, ptc/2013-09-05 (2013)
18. Ochodek, M., Koronowski, K., Matysiak, A., Miklosik, P., Kopczynska, S.: Sketching use-case scenarios based on use-case goals and patterns. In: Proceedings of XVIIIth KKIO Software Engineering Conference on Software Engineering: Challenges and Solutions, pp. 17–30 (2017). https://doi.org/10.1007/978-3-319-43606-7_2
19. Overgaard, G., Palmkvist, K.: Use Cases: Patterns and Blueprints. Addison Wesley, Reading (2005)
20. Powell, S.G., Baker, K.R., Lawson, B.: Errors in operational spreadsheets. J. Organ. End User Comput. **21**(3), 24–36 (2009). https://doi.org/10.4018/joeuc.2009070102
21. Simons, A.J.H.: Use cases considered harmful. In: Proceedings of the 29th Conference on Technology of Object-Oriented Languages and Systems, Nancy, France, pp. 194–203. IEEE Computer Society Press, June 1999. https://doi.org/10.1109/tools.1999.779012
22. Śmiałek, M., Jarzebowski, N., Nowakowski, W.: Runtime semantics of use case stories. In: 2012 IEEE Symposium on Visual Languages and Human-Centric Computing (VL/HCC), pp. 159–162. IEEE (2012). https://doi.org/10.1109/VLHCC.2012.6344506

23. Śmiałek, M., Jarzebowski, N., Nowakowski, W.: Translation of use case scenarios to Java code. Comput. Sci. **13**(4), 35–52 (2012). https://doi.org/10.7494/csci.2012.13.4.35

24. Śmiałek, M., Nowakowski, W.: From Requirements to Java in a Snap: Model-Driven Requirements Engineering in Practice. Springer, Cham (2015). ISBN 978-3-319-12837-5

25. Śmiałek, M., Straszak, T.: Facilitating transition from requirements to code with the ReDSeeDS tool. In: 2012 20th IEEE International Requirements Engineering Conference (RE), pp. 321–322. IEEE (2012). https://doi.org/10.1109/RE.2012.6345825

Software Modelling and Construction

On the Functional Specification
of Queries in OCL

Piotr Kosiuczenko[✉]

Institute of Information Systems, WAT, Warsaw, Poland
piotr.kosiuczenko@wat.edu.pl

Abstract. Query operations are used to evaluate constraints such as
invariants, pre- and post-conditions. The UML standard requires that
a query is pure; i.e. it must not change the global system state. However, operations used for checking validity of conditions quite often have
side effects. As a result, the global system state changes, the problem
is that the validity of constraints should not be changed. Purity is a
subject of intensive research; unfortunately this notion was not properly
investigated in the case of OCL. This paper investigates how the very
restrictive notion of purity can be relaxed in order to be applicable in
the case of state changing operations. It is shown how to define queries
in functional terms relatively to views. Relative queries are classified into
rigid- and non-rigid ones. It is shown that non-rigid queries preserve pre-
and post-conditions; whereas rigid queries preserve also invariants.

Keywords: UML · OCL · Query · Purity · Formal semantics

1 Introduction

Contracts are the prevailing way of specifying object-oriented systems from the
client point of view (see [15,16]). They assign responsibilities to client/caller and
to system implementer/callee. In the realm of UML [22], contracts are expressed
in Object Constraint Language (OCL) [20] (see also [4,25]). The Model-Driven
Architecture (MDA) is a software design approach, which supports model-driven
engineering of software systems [19] (see also [3,24]). It provides a very powerful notion of view, which allows one for a modular system specification. The
client/caller is responsible for satisfaction of the pre-condition before the corresponding method is called. Usually the validity of a pre-condition is checked
by executing queries. The client executes get-methods to check the state of the
underlying system.

The concept of queries, i.e. methods without side effects, stems from
Meyer [16]. Such methods are called pure. In the context of Java Modeling
Language, purity is a subject of active research (cf. e.g. [7,17,18]). It has been
observed that in many cases, the purity assumption is too strong [14]. Therefore a weaker notion of purity was proposed to allow side effects effects such as
updates to caches and security of information (cf. [2]). It has been also observed

© Springer Nature Switzerland AG 2019
P. Kosiuczenko and Z. Zieliński (Eds.): KKIO 2018, AISC 830, pp. 53–68, 2019.
https://doi.org/10.1007/978-3-319-99617-2_4

that it is possible to define boolean conditions with benevolent side effects in if then else statements [2].

In the context of UML and OCL, the concept of purity and query have not been properly investigated. In UML, the notion of query is defined somewhat imprecisely as an operation, which does not change the global system state [22]. This requirement is very restrictive, since it is very common to monitor all activities of a client. In that case, even get-methods are not queries in the strict UML sense, since their use is monitored and recorded.

Above mentioned issue may be seen as a kind of uncertainty principle, because experiments/observations made by a client in terms of query execution may influence the system state. If all actions are monitored, then there are no real queries. This poses the problem of formula validation. An execution of a state changing method can change the validity of a pre-condition or an invariant. For example, a pre-condition can be valid before a get-method execution, and no more valid after the execution. The situation is even worse, if aspect-oriented programming is used. Aspect-oriented languages such as AspectJ (cf. e.g. [11]) allow catching every method call and every access to an attribute. An advice allows then to take appropriate action. Consequently, aspect-oriented programming makes the use of design by contract hard, if possible at all.

In this paper, we investigate what it really means for an operation to be a query and in which circumstances the UML requirements imposed on queries can be relaxed. Since the concept of query plays an important role, it should be as general as possible. We argue that queries can be defined relatively to views using a recently proposed extension of OCL [12]. We also investigate formal properties of queries. In general, the monitoring of client's steps is unproblematic if it does not change validity of the contract. Above mentioned problems can be remedied by the concept of relative query, i.e. query defined relatively to a view. If a contract is expressed in terms of a view, then the execution of a relative query must not change the observable system state. Consequently, execution of a relative query must not change validity of an operation's pre-condition, and in general it must not change values of terms.

We introduce the notion of local state-indistinguishability and a stronger notion of state-indistinguishability. We distinguish between rigid queries, which must not create new objects, and non-rigid queries, which may result in the creation of new objects. We show that non-rigid queries defined relatively to a view result in locally indistinguishable states and that they preserve the corresponding pre-conditions if they are expressed in terms of the underlying view and if they do not contain the predefined feature *allInstances* (this feature defines all instances of the underlying class). the corresponding class (cf. e.g. [10]).

The paper is organized as follows. In Sect. 2, we discuss the query definition in UML and the way relative queries can be defined. In Sect. 3, we present formal semantics of queries, which is a slight modification of the semantics proposed in [9]. We prove that relative queries preserve the validity of formulas expressed in terms of the corresponding view. Section 4 concludes this paper.

2   Query Specification

In this section we discuss the definition of queries in UML. We investigate how to extend the notion of query in order to be applicable to state changing operations. We show how to define relative queries using an extension of OCL and present an example.

2.1   Query Definition in UML

Queries play an important role in UML and OCL. They are used in contracts. They can be executed by a system's client to figure out validity of pre-conditions and invariants. The OCL standard requires that OCL expressions are guaranteed to be without side effect. When an OCL expression is evaluated, it simply returns a value, but the evaluation cannot change anything in the global system state, even though an OCL expression can be used to specify a state change (see [20], Sect. 9.6.3.1).

The UML standard (Sect 9.6.4 of [22]) defines a query to be an operation without side effects. Such an operation doesn't change the global system state. An operation is a query, if in the metamodel the attribute *isQuery* has the value *true*. UML expresses this in the following way (cf. [22], Sect. 7.6.3.1): "If the *isQuery* property is true, an invocation of the Operation shall not modify the state of the instance or any other element in the model.", Furthermore the standard specifies (see Subsect. 9.11.4): "*isQuery : Boolean*[1..1] = *false* specifies whether an execution of the *BehavioralFeature* leaves the state of the system unchanged (*isQuery* = *true*) or whether side effects may occur (*isQuery* = *false*)".

The UML standard requires also that: "In a class model, an operation or method is defined to be side effect free if the isQuery attribute of the operations is true." (cf. [22], 7.5). Similarly, an evaluation of a constraint must not have side effects (cf. [22], Sect. 7.3.10). The previous UML 2.1 standard [21] justifies those requirements by saying that "avoiding side effects in tests will greatly reduce the chance of logical errors and race conditions in a model and in any code generated from it".

The reason why those restrictions are made is that query execution must not make a valid formula invalid or an invalid formula valid. In the Design by Contract approach (DbC), clients/callers are responsible for ensuring that an operation's pre-condition is satisfied before the operation is executed. The validity of a constraint must not changed during its evaluation. In the other case, the client may wrongly conclude for example that the pre-condition is valid and execute the operation in a state in which the pre-condition is no more valid. The operation execution may then result in an inconsistent state, or return an incorrect value. However, it is very common to monitor all client actions. Get-methods often have side effects on the system state. We need to define the concept of query so that on the one hand it would allow changing system state, but on the other hand would guarantee that validity of formulas is not changed.

In the next section we show how to define contract preserving queries using a recently proposed extension of OCL [12,13].

2.2 Definition of Relative Queries in Extended OCL

As explained in the previous section, we need to define queries in such a way that on the one hand it would allow object creation and logging of client's actions, but on the other hand it would make sure that execution of queries does not change validity of a contract. It is intuitively clear that non-invasive observations do not interfere with contracts. The question is under what conditions contracts are preserved by an operation execution and what is the meaning of formulas containing operations changing global system states. To answer this question we distinguish between relative queries, which are defined relatively to a view, and absolute ones. We distinguish also between rigid queries and non-rigid ones. Non-rigid queries may create objects in the corresponding view; whereas rigid queries must not create such objects.

A view is a set of model elements (cf. [12,19]). It can be defined by a package, a class diagram and in general by an OCL constraint [12]. We say that an OCL term F is defined in terms of a view, if for every property such as attribute, association-end and *allInstances* feature occurring in F, the property belongs to the view (see Subsect. 3.2 for a formal definition). We treat formulas as boolean-valued terms. Let *my_view* be a view. An operation f is a rigid relative query defined relatively to *my_view*, if it can be defined by a formula of the form:

context C::f : T
pre : *Pre*
post : *result = F and Cond*
in *my_view* **modifies only** : **nothing**

where the pre-condition *Pre* and the term F are defined in terms of the view. We allow the term F to contain the variable *result*, but *Cond* must not contain it. The formula *Cond* specifies the side effect of f. This OCL-constraint contains a modifies-clause, which is not present in the standard OCL (see [12,13] for a formal definition). This kind of clause is interpreted as a conjunction of invariability formulas. For every object-attribute/association-end a defined by *my_view* and for the corresponding class A, the conjunction contains a formula of the form:

   $A.allInstances$–>$forAll(o \mid not(o.oclIsNew())$ *implies* $o.a = o.a@pre)$

Similarly, if *my_view* includes a class-attribute $C.c$, or the predefined feature $C.allInstances$, then the conjunction contains the formula $C.c = C.c@pre$, or the formula $C.allInstances = C.allInstances@pre$, respectively. Let *inv_conj* be the resulting conjunction. The constraint above can be seen as an abbreviation of the following constraint:

context C::f : T
pre : *Pre*
post : *result = F and Cond and inv_conj*

inv_conj restricts the side effect. It prohibits the modification of those attributes and association-ends which belong to the view. It prohibits also creation and deletion of objects of a class C, if $C.allInstances$ belongs to the view. However, it does not prohibit modification of those attributes and association-ends which do not belong to the view. Similarly, it does not prohibit creation or deletion of new objects of a class C if $C.allInstances$ does not belong to the view.

The inclusion of the predefined feature $C.allInstances$ into a view makes this feature observable. In a sense, it means that a client can access all objects of the class C and observe their creation and deletion. However it is rather uncommon for a client to have direct access to all objects. Usually a client can access objects only by executing methods and evaluating attributes. It is possible to skip this feature in a invariability clauses. We call an invariability clause non-rigid, if it has the form:

in *my_view* **modifies** : **nothing**

The semantics of this clause is equal to the semantics of the following clause:

in *restricted_view* **modifies only** : **nothing**

Where *restricted_view* is obtained from *my_view* by removing all predefined features of the form $C.allInstances$. In this case, creation and deletion of objects from arbitrary classes is allowed as long as associations and attributes defined by this view are not modified. We call q a non-rigid relative query if it is defined by a non-rigid modifies-clause of the form:

context $C::q$: T
pre : *Pre*
post : *result* $= F$ and *Cond*
in *my_view* **modifies** : **nothing**

where *Pre*, *F* and *Cond* are defined in terms of *my_view* and *Cond* does not contain the variable *result*. The query q is absolute and non-rigid if *my_view* contains all model elements included in the underlying model. A query in the sense of UML is a query which is absolute and rigid.

2.3   Example

In this subsection, we present a specification of lists. We show how to define queries relatively to a view so that they preserve contract validity. The class diagram in Fig. 1 shows a list structure. A list is composed of an anchor object of class *List* and a number of *ListElement*-objects. The first element of a list is referenced by the private association-end *first*. A list element references the next element by the private association-end *next*. There are public operations *getFirst* and *getNext* to access objects referenced by those associations. Those two operations have side effects. Their execution is recorded in the static attribute h of class *History*. Whenever an operation is executed, a new object of class *HistItem* is created and appended at the end of the sequence h. The operation *getFirst*()

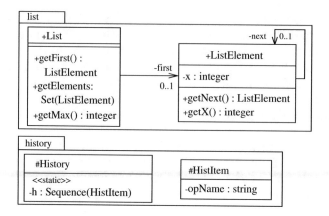

Fig. 1. List with an Anchor.

is specified as follows:

> **context** *List::getFirst() : ListElement*
> **post** : *result = self.first and*
> *History.h–>exist(hi | hi.oclIsNew() and*
>   *hi.opName = "getFirst" and h = h@pre–>including(hi))*
> **in** *list* **modifies** : **nothing**
> **in** *history* **modifies** *mod_pH* : *History.h*

Observe that *getFirst* is an absolute, non-rigid query. The invariability conjunction corresponding to this query has the form (see Subsect. 2.2):

> *List.allInstances–>forAll(l | not(l.oclIsNew()) implies l.first = l.first@pre)*
> *and ListElement.allInstances–>forAll(el | not(el.oclIsNew()) implies*
>   *el.next = el.next@pre and el.x = el.x@pre)*
> *and HistItem.allInstances–>forAll(hi | not(hi.oclIsNew()) implies*
>   *hi.opName = hi.opName@pre)*

Note that *getFirst* is not a rigid query, since, according to the semantics of invariability clauses, new objects of class *List*, *ListElement* and *HistItem* may be created as a side effect. The operation *getNext* can be specified in a similar way.

The collection of elements contained in a list *self.elements* is defined with the help of the auxiliary function *successorsOf*, which collects all successors of a given list element *el* in the set *Acc*:

> **context** *List* **def** :
> *self.elements* : *Set(ListElement) = if self.first = null then Set{}*
>   *else self.first.successorsOf(Set{self.first}) endif*

Where the operation **successorsOf** is specified as follows:

context *ListElement* **def** :

$$el.successorsOf(Acc : Set(ListElement)) : Set(ListElement) =$$
$$if\ el.next\ =\ null\ or\ Acc{-}{>}includes(el.next)\ then\ Acc$$
$$else\ el.next.successorsOf(Acc{-}{>}union(Set\{el.next\}))\ endif$$

The operation *getElements* returns all elements of the underlying list. The execution of this operation results in appending a new *HistItem*-object with the attribute *opName* equal to "*getElements*" at the end of h:

context *List::getElements*() : *Set(ListElement)*

post : $result = self.elements$ *and*

$History.allInstances{-}{>}exist(hi \mid hi.oclIsNew()$ *and*

$hi.opName = $ "*getElements*" *and* $h = h@pre{-}{>}including(hi))$

in *list* **modifies** : **nothing**

in *history* **modifies** mod_pH : *History.h*

The operation *getMax* returns the maximal integer contained in a list, if the list is not empty. We skip the specification of the class-attribute h.

context *List::getMax*() : *integer*

pre pre_getMax : $self.elements{-}{>}notEmpty()$

post $post\_getMax$: $self.elements.x{-}{>}includes(result)$ *and*

$self.elements{-}{>}forAll(o \mid o.x <= result)$ *and ...*

in *list* **modifies** : **nothing**

in *history* **modifies** mod_pH : *History.h*

A client can access list's properties only via relative queries, since all attributes and association-ends are private. Consequently, in order to figure out validity of pre-conditions, the client may have to execute those queries. For example, *getMax* can be executed only if the underlying list is not empty. The elements of a list can be accessed only by executing queries such as *getElements*.

The list specification presented above contains monitoring details, which are not relevant for a client using only the list functionality. It is possible to specify the client relevant part without disclosing that execution of operations is monitored. From the client point of view, the get-methods are real queries, since they do not change anything in the corresponding view. Therefore one can express the corresponding constraints in terms of those relative queries. We restrict the above specification to client's view shown in Fig. 2. This view is determined by package *list*. We call it *client's_view*.

context *List::getFirst*() : *ListElement*

post $post\_getFirst$: $result = self.first$

in *client's_view* **modifies** : **nothing**

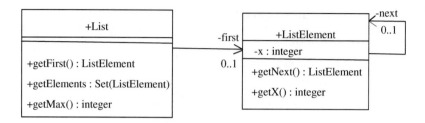

Fig. 2. Client's View.

The operation *getMax* is specified similarly:

context $List::getMax() : integer$
pre : $self.elements{-}{>}notEmpty()$
post : $self.elements.x{-}{>}includes(result)\ and$
$$self.elements.x{-}{>}forAll(y \mid y <= result)$$
in $client's\_view$ **modifies** : **nothing**

The use of invariants is somewhat problematic from the implementation point of view, since they concern all objects of a given class (cf. e.g. [10]). The invariant *inv_pos* concerns all objects of the class *List*. The validation of invariants at runtime is usually not feasible. On the other hand if a pre-condition does not include the feature *allInstances*, then checking its validity is usually computationally inexpensive, since it is enough to evaluate the actual parameters. This is why often invariants are avoided in specifications and only pre- and post-conditions are used. Nevertheless, in some cases invariants are necessary. For example, we may assume that all integers stored in a list are nonnegative:

context $List$ **inv** inv_pos :
$self.elements{-}{>}forAll(o \mid o.x >= 0)$

The question is whether pre-conditions of operations such as *getMax* can be evaluated using above defined queries without changing their validity. We can ask similarly if above defined queries preserve the invariant.

3   Formal Foundations

In this section, we investigate the concept of relative query in formal terms. The first subsection presents an OCL semantics, which is a slight variation of the semantics proposed in [9]; the difference is that we treat UML queries and state changing operations in the same way. We present this semantics as far as it is needed to prove that relative queries preserve contracts. In the second subsection, we define the notion of state-indistinguishability, we define the meaning of terms including relative queries and prove that relative queries preserve contracts defined in terms of the corresponding view. In the third subsection we demonstrate how the results apply to the list example.

3.1 Semantics of OCL

In this subsection we present a formal semantics of OCL. The OCL semantics defined in [5,9] strictly differentiates between UML queries and state-changing operations. Queries are part of the functional signature; whereas non-query operations are treated as relations, which for a given system state output a new system state and optionally a value. In OCL, the state parameter is hidden. It occurs only in formalization of OCL terms. Most formal semantics of OCL interpret classes as sets. For example, the semantics proposed in [9] interprets $C.allInstances$ as the set of all objects of class C, which exist at a given moment of time. In that semantics, one can define $C.allInstances@pre$ as the set of objects existing at the moment when a method is invoked. The set of objects, which exist before and after method execution, can be defined as the intersection of $C.allInstances@pre$ and $C.allInstances$.

The semantics used in this paper is a slight modification of the semantics defined in [9]. We use here also a bit different notation. We do not dwell on the model theoretic foundations, as they are fairly basic and can be found in any book on model theory (cf. e.g. [6]). A class signature (CS) Σ has the form $(S, \leqslant, F)$. S is a set of sorts. It includes sorts for classes, types such as $Set(C)$, and for basic OCL types such as $Integer$. We assume that there is a sort $State$ corresponding to global system states. We assume that the set of sorts is partially ordered by the relation $\leqslant$. F is a set of typed function symbols corresponding to attributes, associations, operations, and to predefined OCL functions such as $includes$. We use the prefix notation in CS-function symbols formalizing OCL terms. For example, if a is an attribute of class C with value of type T, then CS contains the function symbol $a\_\_ : State \times C \to T$. Note that the additional parameter $State$ corresponds to the current state of the underlying system. The value of attribute a can change as the system evolves. Consequently, its value depends on the state and the underlying object of class C. Similarly for a class T, the pre-defined feature $T.allInstances$ is formalized by the function $T.allInstances\_ : State \to Set(T)$. The predefined function $includes$ is formalized by the function $includes\_\_ : Set(T) \times T \to Boolean$. Note that we do not need to know the state to figure out whether an object belongs to a set or not. Therefore we say that $includes$ is a state independent function.

Let $\mathbb{M}$ be a model corresponding to the class signature Σ (we will say that $\mathbb{M}$ is a CS-model). For an OCL type T, $T^{\mathbb{M}}$ denotes the set corresponding to the sort T. If T is a class, then $T^{\mathbb{M}}$ can be interpreted as the corresponding name/address space. We assume that for every type T different from $State$, the undefined element $\perp$ belongs to $T^{\mathbb{M}}$. For a function symbol f, $f^{\mathbb{M}}$ denotes the interpretation of f in the model $\mathbb{M}$. In OCL every term defines a set of objects and/or values of a basic type, or it is undefined, i.e. returns $\perp$. To make the notation more compact we write $T_{\sigma}^{\mathbb{M}}$ instead of $T.allInstances^{\mathbb{M}}(\sigma)$ to denote the set of objects of class T, which exist in the state σ. Note that for a state σ, $T_{\sigma}^{\mathbb{M}} \subseteq T^{\mathbb{M}}$.

Inheritance corresponds to sub-sort relation and set inclusion. If T_1 subclasses T_2, then we assume that the corresponding sorts are ordered $T_1 \leqslant T_2$, and that the set inclusion $T_1^{\mathbb{M}} \subseteq T_2^{\mathbb{M}}$ holds. Let σ be a system state before an operation execution and let σ' be a system state after the execution. The set of objects of class T, which exist before and after the execution, is equal to $T_\sigma^{\mathbb{M}} \cap T_{\sigma'}^{\mathbb{M}}$. In the case of predefined OCL types such as *Real*, we assume that $Real^{\mathbb{M}}$ is the set of all real numbers and that the set of Reals is immutable in all system states, i.e. the equation $Real_\sigma^{\mathbb{M}} = Real^{\mathbb{M}}$ holds for every state σ (cf. [9]).

For an attribute/association a with argument of class C and value of type T, a state σ, and an object which exists in state σ (i.e. $o \in C_\sigma$), we require that the value exists in this state as well, i.e. $a^{\mathbb{M}}(\sigma, o) \in T_\sigma^{\mathbb{M}}$; we also assume that $a^{\mathbb{M}}(\sigma, \bot) = \bot$. The OCL primitive *oclIsNew* is formalized by a function such that $oclIsNew^{\mathbb{M}}(\sigma, \sigma', o) = true$ iff $o \in T_{\sigma'}^{\mathbb{M}} \setminus T_\sigma^{\mathbb{M}}$. We say that a function $f(x_1 : T_1, ..., x_n : T_n) : T$ is a state independent, if it does not have a state parameter. There are state independent functions such as *includes*, $+$, *not*, *forAll*, *and*, or *if then else end*. The last three functions are non-rigid. We assume that if $f(x_1 : T_1, ..., x_n : T_n) : T$ a state independent operation and $o_i \in T_{i\sigma}^{\mathbb{M}}$, for $i = 1, ..., n$, then $f^{\mathbb{M}}(o_1, ..., o_n) \in T_\sigma^{\mathbb{M}}$.

An operation $op(x_1 : C_1, ..., x_n : C_n) : T$ is formalized by a pair of functions. The first function is of the form $op : State \times C_1 \times ... \times C_n \to T$. The second one is of the form $\overrightarrow{op} : State \times C_1 \times ... \times C_n \to State$; it models state change. The state returned by this function corresponds to the state after operation execution. If the operation op is a query in the strict UML sense, then the state remains unchanged, i.e. $\overrightarrow{op}(s, x_1, ..., x_n) = s$.

We treat the OCL-term *self.a* as an abbreviation of $a(s, self)$, where s is a variable of sort *State*. The value of a class-attribute depends on the corresponding state. Consequently we treat class-attribute of the form $C.c$ as an abbreviation of $C.c(s : state)$. In general, we treat OCL terms as abbreviation of the terms of the class signature; we use prefix notation in the formalization. We treat all formulas as boolean valued terms. A formula is satisfied by a valuation, if the term corresponding to the formula has value *true* for this valuation. Let $t(x_1 : T_1, ..., x_n : T_n)$ be a CS term and let $\mathbf{v} = [x_1 \mapsto o_1, ..., x_n \mapsto o_n]$ be a corresponding valuation, we will often write $t^{\mathbb{M}}\mathbf{v}$ instead of $t^{\mathbb{M}}(o_1, ..., o_n)$. For example the OCL formula *self.a* $<=$ *self.b* is an of the CS-term $a(s, self) < b(s, self)$.

For a CS-term t and a CS-model $\mathbb{M}$, $t^{\mathbb{M}}$ denotes the corresponding function. Let $\mathbf{v} = [s \mapsto \sigma, self \mapsto o]$ be a valuation which maps the variable s to the state σ and the variable *self* to the object o. The above formula is satisfied by $\mathbf{v}$ if, and only if, $a^{\mathbb{M}}(\sigma, o) <^{\mathbb{M}} b^{\mathbb{M}}(\sigma, o)$ has value *true*.

If there are get-operations for attribute a and b which do not change the state of the system, then we can write also *self.getA* $<=$ *self.getB*; this formula can be then formalized in two ways: $getA(s, self) < getB(\overrightarrow{getA}(s, self), self)$, and $getA(\overrightarrow{getA}(s, self), self) < getB(s, self)$. The first formula corresponds to the case when the left hand side is evaluated first. In this case the new state is fed into the right hand side. This formula can be equivalently expressed as follows: $getA(s, self) < getB(s', self)$ *and* $s' = \overrightarrow{getA}(s, self)$. The second formula corresponds to the case when the right hand side is evaluated first. In this paper we investigate under which conditions the value returned by an operation does not depend on a state. Those conditions allow us to skip state returning functions, i.e. functions of the form $\overrightarrow{op}$.

3.2 Formalizing Relative Queries

In this subsection we introduce the notion of relative state-indistinguishability. We differentiate between local indistinguishability and global indistinguishability. We prove that execution of relative non-rigid queries results in locally indistinguishable states, while execution of rigid queries results in indistinguishable states. We show that non-rigid queries preserve pre-conditions if they are defined in terms of the underlying view and do not contain the feature *allInstances*. We show also that rigid queries preserve all pre-conditions and invariants defined in terms of the underlying view.

We say that two states are indistinguishable in respect to a certain view, if for all objects existing in both states, and all attributes and associations belonging to that view, the value of those attributes/associations is the same. Below we treat views as sets. Formally, we say that states σ and σ' are locally indistinguishable in respect to a view *my_view*, if and only if conditions (1) and (2) are satisfied. The states are indistinguishable if in addition the condition (3) is satisfied.

1. For every object-attribute/association a, if $a \in my\_view$ and if C is the class corresponding to the type of the argument of a, then for all $o \in C_\sigma^{\mathrm{M}} \cap C_{\sigma'}^{\mathrm{M}}$ the equation $a^{\mathrm{M}}(\sigma, o) = a^{\mathrm{M}}(\sigma', o)$ holds.
2. If the class-attribute $C.c$ belongs to *my_view*, then $C.c^{\mathrm{M}}(\sigma) = C.c^{\mathrm{M}}(\sigma')$.
3. If the feature $C.allInstances$ belongs to *my_view*, then $C_\sigma^{\mathrm{M}} = C_{\sigma'}^{\mathrm{M}}$.

Note that state-indistinguishability is an equivalence relation, i.e. reflexive, symmetric and transitive, if for every object-attribute/association a and the corresponding class C, $C.allInstances$ belongs to *my_view*.

We assume that there is a number of recursive definitions of the form:
$$f_1(x_1, ..., x_k) = F_1(x_1, ..., x_k), ..., f_l(x_1, ..., x_k) = F_l(x_1, ..., x_k)$$
such that, for $i = 1, ..., l$, f_i is a function symbol and F_i is a term, which does not include state returning function (see Subsect. 3.1). For a term t, let $\mathsf{Symb}(t)$ denotes the set of all symbols formalizing attributes, associations, function-symbols f_i and $C.allInstances$ features occurring in the term t. We define the set of relevant symbols corresponding to t as the smallest set of symbols defined by the equation $\mathsf{RelSymb}(t) = \mathsf{Symb}(t) \cup \bigcup_{f_j \in \mathsf{Symb}(t)} \mathsf{RelSymb}(F_j)$. Let $F_A \subseteq F$ be the set of all function symbols, which formalize association symbols, attribute symbols and $C.allInstances$ features.

The set of relevant attribute and association symbols is defined as follows: $\mathcal{A}(t) = \mathsf{RelSymb}(t) \cap F_A$ (cf. Subsect. 2.2). The set $\mathcal{A}(t)$ contains only attributes, associations and the predefined feature *allInstances*. It is obtained from $\mathsf{RelSymb}(t)$ by removing all recursively defined symbols.

For a type T and a sequence of states $\sigma_1, ..., \sigma_n$, $T_{\sigma_1, ..., \sigma_n}^{\mathrm{M}}$ denotes the set of objects of type T which exist in all those states, i.e. it denotes the intersection $T_{\sigma_1}^{\mathrm{M}} \cap ... \cap T_{\sigma_n}^{\mathrm{M}}$.

The next lemma says that the value of a term t is equal for two valuations, if three conditions are satisfied. The first condition requires that the term t must be defined in terms of the underlying view. The second and third condition require that the two valuations can differ only by assigning different indistinguishable states to state variables. The third condition requires that during an operation execution the actual parameters of the operation cannot be deleted and must be kept unchanged.

Lemma: Let my_view be a view. Let $t(s_1, ..., s_m : State, x_1 : T_1, ..., x_n : T_n) : T$ be a term of signature Σ which does not include state returning functions of the form $\overrightarrow{op}$ and let the sort T be different from the sort $State$. Let $\mathbf{v}$ and $\mathbf{v}'$ be valuations. Moreover, let $\sigma_i = \mathbf{v}(s_i)$ and $\sigma_i' = \mathbf{v}'(s_i)$, for $i = 1, ..., m$. If the following conditions are satisfied:

1. $\mathcal{A}(t) \subseteq my\_view$.
2. For $i, j = 1, ..., m$, the states σ_i, σ_j' are indistinguishable in respect to my_view.
3. For every variable $x_i : T_i$, $\mathbf{v}(x_i) = \mathbf{v}'(x_i) \in T_{i\,\sigma_1,\sigma_1',...,\sigma_n,\sigma_n'}^{\mathbb{M}}$.

Then $t^{\mathbb{M}}\mathbf{v} = t^{\mathbb{M}}\mathbf{v}'$, and $t\mathbf{v}, t\mathbf{v}' \in T_{\sigma_1,\sigma_1',...,\sigma_m,\sigma_m'}^{\mathbb{M}}$.

Proof: Proving theorems for languages as general as OCL is usually not easy. However, in our case there are only two kinds of relevant OCL features: there are state dependent features and state independent features.

The semantics of recursive functions is usually defined using the Kleene fixed-point theorem. This theorem says that every continuous function defined on a complete partially ordered set has the least fixed-point. The monotone function is usually defined by unwinding of the recursively defined function symbol, i.e. by an application of recursive definitions (cf. e.g. [1]). Let $\mathbb{M}^{f \mapsto \perp}$ be a model of signature Σ, which is obtained from $\mathbb{M}$ by assigning the undefined value $\perp$ to the recursively defined function symbols f_i. For a valuation $\mathbf{v}$, a function $f_i^{\mathbb{M}}$ has value a different from $\perp$ if and only if there is an unwinding v of f_i, such that $v^{\mathbb{M}^{f \mapsto \perp}}\mathbf{v} = a$.

Let u be an unwinding of t. Note that it is enough to prove that conditions (1), (2) and (3) imply that
$$u^{\mathbb{M}^{f \mapsto \perp}}\mathbf{v} = u^{\mathbb{M}^{f \mapsto \perp}}\mathbf{v}' \text{ and } u\mathbf{v}, u\mathbf{v}' \in T_{\sigma_1,\sigma_1',...,\sigma_m,\sigma_m'}^{\mathbb{M}^{f \mapsto \perp}}.$$
Note that if u is of minimal hight, then since its type is different from $State$ and since it does not include state returning function symbols, u is either a variable, or class-attribute of the form $C.c$, or a predefined feature of the form $C.allInstances$. If u is a variable, then $u = x_i$ for some i, since T is different from $State$. From condition (3) follows that $\mathbf{v}(x_i) = \mathbf{v}'(x_i)$. If u has the form $C.allInstances(s_i)$ or $C.c(s_i)$, then the property follows from the definition of indistinguishability.

Assume that the property holds for all terms of hight smaller than the hight of u. Let u have the form $a(s_j, v) : T$ for an attribute/association a and a term v of type C. From the inductive assumption follows that

$v^{M^{f \mapsto \perp}} \mathbf{v} = v^{M^g} \mathbf{v}' \in C^{M^{f \mapsto \perp}}_{\sigma_1, \sigma_1', ..., \sigma_m, \sigma_m'}$. From the indistinguishability property follows that for all $o \in C^{M^{f \mapsto \perp}}_{\sigma_1, \sigma_1', ..., \sigma_m, \sigma_m'}$, the equation $a^{M^{f \mapsto \perp}}(\sigma, o) = a^{M^{f \mapsto \perp}}(\sigma, o)$ holds, and moreover $a^{M^{f \mapsto \perp}}(\sigma, o) \in C^{M^{f \mapsto \perp}}_{\sigma_1, \sigma_1', ..., \sigma_m, \sigma_m'}$ (see Subsect. 3.1). In particular these properties hold for every $o \in v^{M^{f \mapsto \perp}} \mathbf{v} = v^{M^{f \mapsto \perp}} \mathbf{v}'$. Consequently, $u^{M^{f \mapsto \perp}} \mathbf{v} = u^{M^{f \mapsto \perp}} \mathbf{v}' \in C^{M^{f \mapsto \perp}}_{\sigma_1, \sigma_1', ..., \sigma_m, \sigma_m'}$.

If u has the form $f_i(t_1, ..., t_k)$, for a recursively defined symbol f_i, then u is equal to $\perp$ in $M^{f \mapsto \perp}$ for all states and all arguments.

Let $u(x_1, ..., x_n)$ have the form $f(t_1 : T_1, ..., t_k : T_k) : T$, for a state independent function f. Then $f(t_1, ..., t_k)^{M^{f \mapsto \perp}} \mathbf{v} = f^{M^{f \mapsto \perp}}(t_1^{M^{f \mapsto \perp}} \mathbf{v}, ..., t_k^{M^{f \mapsto \perp}} \mathbf{v}) = f^{M^{f \mapsto \perp}}(t_1^{M^{f \mapsto \perp}} \mathbf{v}', ..., t_k^{M^{f \mapsto \perp}} \mathbf{v}') = f(t_1, ..., t_k)^{M^{f \mapsto \perp}} \mathbf{v}' \in T_{\sigma_1, \sigma_1', ..., \sigma_m, \sigma_m'}$ (see Subsect. 3.1).   ♦

Consider a constraint of the form:

context $C::op(x_1, ..., x_m)$
pre : $Pre(self, x_1, ..., x_m)$
post : $Post(self, x_1, ..., x_m)$

A contract allows execution of op only if the pre-condition Pre is satisfied. Suppose that a client is going to execute operation op with actual parameters $o, p_1, ..., p_m$. Since the client is responsible for the validity of the pre-condition, Pre has to be checked. In order to do that, the client may execute a number of relative queries. Those actual parameters are kept by the client during the execution. If the evaluation shows that the pre-condition is satisfied, then the client executes op with those parameters. The value of $Pre(o, p_1, ..., p_m)$ must not change during the execution of relative queries. Observe that if any of the actual parameters is deleted, then the operation cannot be executed anymore with those parameters. Obviously it does not make sense to require that the contract is still valid for parameters which do not exist anymore.

Let Pre be defined in terms of *my_view*, i.e. $\mathcal{A}(Pre) \subseteq my\_view$. From the lemma proved above follows that if Pre is satisfied by a valuation $\mathbf{v}$ of the form $[self \mapsto o, x_1 \mapsto p_1, ..., x_m \mapsto p_m]$ in a state σ and if σ' is indistinguishable from σ, then Pre is satisfied by $\mathbf{v}$ in the state σ' too. The invariability clause of a rigid query (see Subsect. 2.2) implies conditions (1), (2) and (3) of the definition of indistinguishability. This implies that the execution of a rigid query defined relatively to *my_view* results in an indistinguishable state and consequently rigid queries preserve Pre.

Pre-conditions usually restrict parameters of the corresponding operation and do not include the feature *allInstances*. Let *restricted_view* be obtained from *my_view* by removing all features of the form $C.allInstances$ and let q be a non-rigid query defined relatively to *my_view*. If the pre-condition Pre does not include the feature *allIstances*, i.e. $\mathcal{A}(q) \subseteq restricted\_view$, then from the lemma above follows that q preserves the validity of Pre.

An invariant concerns all instances of a the corresponding class (cf. e.g. [10]). The lemma implies that a rigid relative query preserves an invariant Inv concerning a class C, if the invariant is defined in terms of the corresponding view, and if the view contains $C.allInstances$.

Formally, let $self.I$ denote an invariant I containing the free variable $self$. The implicit variable $self$ ranges over all objects of the underlying class. If an operation creates a new object of the corresponding class, then the object may violate the invariant $self.I$. Note that we treat formulas as boolean valued terms. From the lemma we can derive the following

Corollary: The validity of a pre-condition Pre is preserved by rigid queries defined relatively to my_view, if $\mathcal{A}(Pre) \subseteq my\_view$. Moreover, if $\mathcal{A}(Pre)$ does not include the feature $allInstances$, then its validity is preserved by non-rigid queries defined relatively to my_view. Moreover, for every class C and every invariant of the form $self.I$ defined in the context of C, if $C.allInstances$ belongs to my_view and $\mathcal{A}(I) \subseteq my\_view$, then the validity of I is preserved by rigid queries defined relatively to my_view.

Note that, an execution of a non-rigid query may violate an invariant, since the query may create a new object inconsistent with the invariant. Similarly, an execution of a non-rigid query may violate a pre-condition Pre containing the predefined feature $C.allInstances$.

3.3   Example Continued

In this subsection we show how the formal concepts developed in the previous subsection can be applied to the list example (see Subsect. 2.3). In that example we identify a client's view corresponding to the list functionality. It abstracts from model elements needed for the monitoring of operations. $client's\_view$ contains association-ends $first$, $next$ and the attribute x. The set of attributes, association-ends and $allInstances$ features included in this view has the form $\{first, next, x, ListElement.allInstances, List.allInstances\}$.

The operation $getMax$ returns the largest number stored in a list. This operation is applicable only if the underlying list is not empty. Checking the corresponding pre-condition requires execution of the relative query $getElements$. Execution of this query changes the global state, but from the client perspective the resulting state cannot be distinguished from the initial one, as those states differ only in respect to the class attribute h. More precisely, in the resulting state, the value of class-attribute h is changed by appending a new object of class $HistItem$. All objects corresponding to the classes $List$ and $ListElement$ remain unchanged. Similarly, the operation $getFirst$ is a non-strict relative query.

The set $elements$ is defined recursively. The corresponding set of relevant function symbols, attributes and association-ends has the form:
$RelSymb(elements) = \{first, next, elements, successorsOf\}$.
The set of the corresponding properties has the form $\mathcal{A}(elements) = \{first, next\}$.

The pre-condition pre_getMax is defined in terms of $first$ and $next$, i.e. $\mathcal{A}(pre\_getMax) = \{first, next\}$. Consequently, this pre-condition is defined in terms of $client's\_view$ and its validity is preserved by $getElements$.

Note that the invariability-formula corresponding to *getElements* is identical with the invariability formula corresponding to *getFirst*. It does not prohibit the creation of new *List-* and *ListElement-*objects. The creation of new lists and in particular new list elements may violate the invariant *inv_pos* as the attribute x may be negative in newly created objects. If *getFirst*, *getElements* were defined using a clause of the form: **in** *client's_view* **modifies only** : **nothing** then they would be rigid queries. The corresponding invariability-formula would include the invariability-formula defined in Subsect. 2.3 and also the following equations:

ListElement.allInstances = ListElement.allInstances@pre

List.allInstances = List.allInstances@pre

Consequently, the validity of the invariant *inv_pos* would be preserved by those queries, since no new objects of classes *List* and *ListElement* would be created and the value of *self.x* would remain unchanged for all already existing objects as specified by the invariability conjunction.

4   Conclusion

The query definition proposed in the UML and OCL standards exclude any change to the system state. In many cases, this assumption is too rigid to be applicable. In this paper, we demonstrated how to define queries using a recently proposed extension of OCL. We demonstrated also how to deal with state changing operations. We identified two classes of queries and proved that non-rigid queries preserve pre-conditions and rigid queries preserve all formulas defined in terms of a view.

References

1. Abramsky, S., Jung, A.: Domain Theory. In: Handbook of Logic in Computer Science, vol. 3, pp. 1–168. Clarendon Press (1994)
2. Barnett, M., Naumann, D., A., Schulte, W., Sun, Q.: Allowing state changes in specifications. In: International Conference on Emerging Trends in Information and Communication Security, ETRICS. LNCS, vol. 3995, pp. 321–336. Springer (2006)
3. Brambilla, M., Cabot, J., Wimmer, M.: Model-driven Software Engineering in Practice. Morgan and Claypool Publishers (2012)
4. Cabot, J. Gogolla, M.: Object constraint language (OCL): a definitive guide. In: Formal Methods for Model-Driven Engineering, SFM 2012, Bertinoro, Italy, 18-23 June 2012. LNCS, Vol. 7320, pp. 58–90. Springer (2012)
5. Cengarle, M., Knapp, A.: OCL 1.4/1.5 vs. OCL 2.0 expressions: formal semantics and expressiveness. Softw. Syst. Model. **3**(1), 9–30 (2004)
6. Chang, C., Keisler J.: Model Theory. North-Holland, rev. edn. (1990)
7. Darvans, Á., Müller, P.: Reasonong about Method Calls in JML. In: Formal Techniques for Java-Like Programms, FTfJLP 2005 (2005)
8. Harel, D.: Statecharts: a visual formalism for complex systems. Sci. Comput. Program. **8**, 231–274 (1987)
9. Hennicker, R., Knapp, A., Baumeister, H.: Semantics of OCL operation specifications. ENTCS **102**(2), 111–132 (2004)

10. Hennicker, R., Baumeister, H., Knapp, Al., Wirsing, M.: Specifying component invariants with OCL. GI Jahrestagung 2001, pp. 600–607 (2001)
11. Kiczales, G., et al.: An overview of AspectJ. LNCS, vol. 2072, pp. 327–355. Springer (2001)
12. Kosiuczenko, P.: Specification of invariability in OCL. In: Nierstrasz, O., et al. (eds.) MoDELS 2006. LNCS, vol. 4199, pp. 676–691. Springer (2006)
13. Kosiuczenko, P.: 17. Specification of Invariability in OCL. J. Softw. Syst. Model. **12**(2), 415–434 (2013)
14. Leavens, G., Yoonsik, C., Clifton, C., Ruby, C., Cok, D.: How the design of JML accomodates both runtime assertion checking and formal verification. Sci. Comput. Program. **55**(1–3), 185–208 (2005)
15. Mitchell, R., McKim, J.: Design by contract by example. Addison-Wesley (2001)
16. Meyer, B.: Object-Oriented Software Construction. Prentice Hall, NJ (1998)
17. Naumann, D.: Observational purity and encapsulation. In: Cerioli, M., (ed.) Fundamantal Aspects of Software Engineering (FASE 2005). LNCS, vol. 3442, pp. 190–204 (2005)
18. Naumann, D.: Observational purity and encapsulation. Theor. Comput. Sci. **376**(3), 205–224 (2007)
19. OMG, MDA Guide, Version 1.0.1, June 2003
20. OMG, OCL Specification, Version 2.4, February 2014
21. OMG, Unified Modeling Language Specification: Superstructure, Version 2.1.1, 05 February 2007 (2007)
22. OMG, Unified Modeling Language Specification: Superstructure, 2.5.1, December 2017
23. Pnueli, A., Shalev, M.: What is in a step: on the semantics of Statecharts. In: Meyer, R., Ito, T. (eds.) TACS 1991. LNCS, vol. 526, pp. 244–264. Springer, New York (1991)
24. Rumpe, B.: Agile Modeling with UML - Code Generation, Testing, Refactoring. Springer (2017)
25. Warmer, J., Kleppe, A.: Object constraint language: getting your models ready for MDA. Addison Wesley (2003)

Multi-level Security System Verification Based on the Model

Andrzej Stasiak[(✉)] and Zbigniew Zieliński

Faculty of Cybernetics, Military University of Technology, gen. Urbanowicza 2,
00-908 Warsaw, Poland
{andrzej.stasiak,zbigniew.zielinski}@wat.edu.pl

Abstract. In the paper the approach to multi-level security (MLS) systems verification on the base of Bell-LaPadula and Biba models is presented. The essence of the proposed approach to analyze properties of MLS security-design models and their instances is models integration and their evaluation and simulation. Properties of the security policy model are expressed as constrains in OCL language. Also, "separability" problem of different security domains is formulated and a method for its verification is proposed. The feasibility of the proposed approach by applying it to the example MLS project is demonstrated.

Keywords: Multi-level security · System verification · Bell-LaPadula model
Biba model

1 Introduction

Developing of dependable Specialized Computer Systems (SCS), that process data with different levels of sensitivity is particularly important for governmental, military or financial institutions. The problem of multi-level security (MLS) systems design has been extensively studied since the early 70s of the twentieth century [1–3]. Various multilevel security models (MLS) have been created to enforce confidentiality and integrity of data. Some of the more popular models are Bell-LaPadula (BLP) Model [1, 2], Biba Model [3], Lipner's Integrity Matrix Model and Clark-Wilson Model [4]. To ensure the confidentiality and integrity of information models of BLP and Biba are frequently used, which provide mandatory access control (MAC) entity (called subject) to the resource (called object). Bell and LaPadula developed lattice-based access control models to deal with information flow in computer systems. Information flow also applies to integrity (Biba) to some extent. Lattice-based access control is one of the essential ingredients of computer security [5].

In the MLS (multi-level security) systems security requirements introduce not only quality characteristics but also constraints under which the system must operate (see §4). Ignoring such constraints during the development process could lead to serious system vulnerabilities.

The basic idea of integrating system design models (expressed usually in UML) with security considerations is not new [6–13]. Such integrated models with both a concrete notation and abstract syntax are called security-design models [8]. In the

© Springer Nature Switzerland AG 2019
P. Kosiuczenko and Z. Zieliński (Eds.): KKIO 2018, AISC 830, pp. 69–85, 2019.
https://doi.org/10.1007/978-3-319-99617-2_5

A. Stasiak and Z. Zieliński

work [8] the security modeling language, called SecureUML, was presented, which is closely related to Role Based Access Control (RBAC). Subsequently, in [9] was shown that security properties of security-design models could be expressed as formulas in OCL and an expressive language for formalizing queries concerning RBAC policies was also proposed. The idea of formulating OCL queries on access control policies was introduced in [10–12], who first explored the use of OCL for querying RBAC policies. However, RBAC has several limitations and its use as a base for modeling security policy of MLS systems is impractical.

In [7] we have proposed a method of software design of MLS-type systems called MDmls, which is based on MDD (Model Driven Development) approach [13]. The essence of the MDmls method is integration of the MLS security models with the system design models expressed in UML-based language [14, 15]. In this work new MlsML profile was elaborated for the security policy verification of MLS systems and incorporates discretionary and mandatory access control on the base of properties of the lattice models (mainly Bell-LaPadula and Biba).

The construction of SCS MLS systems can be based on centralized or distributed computer system. One of the possible approach to the construction of a centralized (i.e., no distributed) computer system with multi-level security is to develop software in the virtualization technology [14, 15] for the separation of independent security domains. Such approach has become today entirely possible thanks to the availability of solutions with virtualization hardware support in modern Intel and AMD processors. Now widely used are the extensions of the x86 architecture, designed to support hardware virtualization as Intel Virtualization Technology in particular VTx, VTd for x86 processors.

In this kind of systems "secure isolation" between the security domains of a shared resources of computer system is needed. Similar problems we could observe in the cloud infrastructure with multi-level security. One of the significant problems in the development of MLS systems is to prove of "separability". This problem lies at the interface of the hardware and software components. So, we extend the security-design models with the topology models [16] that allow binding of hardware and software components of the MLS system.

We see our contributions as follows. First, we extended the approach proposed in [15] to the MLS security models, which are based on the MLS lattice models. Second, we demonstrate, how to formulate OCL expressions to check properties of the MLS lattice models. Next, we proposed the way of integration of the MLS models with a topology model [16] that allow binding of hardware and software components of the MLS system. Finally, we show the feasibility of this approach by applying it to a non-trivial example: MLS security policy and security-design models verification of the Secure Workstation for Special Application (SWSA) Project[1] with the use of IBM RSA tool.

The rest of the work is organized as follows. In Sect. 2 we describe our general approach to security modeling in MLS systems. In Sect. 3 we propose MlsML profile for the confidentiality and integrity verification with BLP and Biba models accordingly.

[1] Project No. OR00014011 supported by The Polish National Center for Research and Development.

In Sect. 4 we describe an example of MLS security policy and MLS security-design models verification. In Sect. 5 we draw conclusions and discuss future work.

2 General Approach to Security Modeling in MLS Systems

In this section we explain our approach to analyze properties of security-design models of MLS systems on the base of the BLP or Biba models and the use of the evaluation and simulation of the models. We propose MlsML language, which in current implementation enables creating MLS lattice models, through which it is possible to study the effects of different policies.

2.1 Problem Statement

Secure software design of the MLS systems employs dedicated tools to verify the confidentiality and the integrity of data using UML models. In general, the UML security models could be embedded in and simulated with the system architecture models, thus the security problems in MLS system can be detected early during the software design.

As was shown in [9], precise analysis of UML models (or their instances) depicted by some diagrams requires definition of the formal semantics of diagrams, that is, definition of an interpretation function $I(\cdot)$, which associates formal (mathematical) structures $I(M)$ to well-formed diagrams M. In general, given a security modeling language with a formal semantics, one can reason about models by reasoning about their semantics. In the case of a MLS security modeling language $\mathcal{L}_{MLS}$, a security model (or a security model instance) M has a property P (expressed as a formula in some logical language) if and only if $M \Rightarrow P$.

We are formulating the problem as follows. Because OCL (Object Constraint Language) is the natural choice for querying UML models, it can be used to constrain and query MLS security-design models. Our approach to analyze properties of MLS security-design models and their instances reduces deduction to evaluation and simulation.

The way of formally analyzing MLS security policies concerning confidentiality and integrity of a designed system and modeled by model M is to express the desired properties of M as OCL queries and evaluate these queries on the UML models or model instances. We investigate the way of automatically evaluating properties of MLS system, in the context of the metamodel of the proposed security-design language MlsML.

2.2 MLS Models

The Bell-LaPadula model focuses on data confidentiality and controlled access to classified information. In this formal MLS model, the entities in an information system are divided into subjects and objects, all subjects and objects are labeled with a security level. The levels represent the relative sensitivity of the data and the clearance of the user on whose behalf the subjects are operating. For semantic reasons of model building the security level of subjects and objects will be distinguished.

Let $C = \{c_1, c_2, \ldots, c_L\}$ denote the ordered set of clauses which represent sensitivity of data used in the MLS system, where $c_i \leq c_{i+1}$ for $1 \leq i < L$. Let $IC = \{\theta_1, \theta_2, \ldots, \theta_C\}$ be the set of categories of information processed in the system.

For each subject $s \in S$ we assign a security level $SL(s)$ as a pair $\langle c_s, A_s \rangle$ and for each of object $o \in O$ we assign a security level $SL(o)$ as a pair $\langle c_o, A_o \rangle$, where c_s, $c_o \in C$ and $A_s, A_o \subseteq IC$. Security levels can be compared. It could be noticed that not all pairs of levels are comparable. This leads to the use of the concept of lattice of security levels.

A dominance relationship $dom(s, o)$ may be introduced between subject $s \in S$ with $SL(s) = \langle c_s, A_s \rangle$ and object $o \in O$ with $SL(o) = \langle c_o, A_o \rangle$, if $SL(s) \geq SL(o)$. It can be expressed as the formula:

$$dom(s, o) \Leftrightarrow (c_s \geq c_o) \wedge (A_o \subseteq A_s). \tag{1}$$

The BLP model is based around two main rules: the simple security property and the star property [1, 2]. The simple security property (ss-property) states that a subject $s \in S$ can read an object $o \in O$ if the formula (1) is hold. The simple security property prevents subjects from reading more privileged data. The star property (*-property) states that a subject can write to an object, if subject is dominated by object.

The Bell La-Padula model does not deal with the integrity of data. It is possible for a lower level subject to write to a higher classified object. The Biba model addresses the problem with the star property of the Bell-LaPadula model, which does not restrict a subject from writing to a more trusted object. Similarly to BLP model integrity levels are defined by labels, consisting of two parts: a classification and a set of categories. Each integrity level will be represented as $IL = \langle \alpha, A \rangle$ where IL is the integrity level, α is the classification and A is the subset of categories. Similar to BLP model the integrity levels then form a dominance relationship. The Biba model is actually a family of different policies that can be used depending on the specific of MLS system [3].

2.3   Metamodel

In our approach, in order to determine the system's security of MLS type, we refer to the UML four-layer hierarchy (Fig. 1). In order to simplify the process of constructing a specialized MlsML language, the specification will be determined at the level of M2, which, in essence, is the same metamodel. In the metamodel the key concepts in the domain will be described, which will be mapped to the UML profile stereotypes with developed MlsML language. In this sense, our proposal of MlsML is an extension of the UML language, not its complete specification.

Level M1 would be a model of the field (description of the MLS system security), in which we use elements developed from meta-model, for the construction of MlsML language expressions in the form of diagrams. Each diagram would be created according to the rules of the UML language, enhanced with a new profile property (which is why it is acceptable to refer to the developed diagrams of the field, both for the concepts of language specified in the profile, as well as in the UML language).

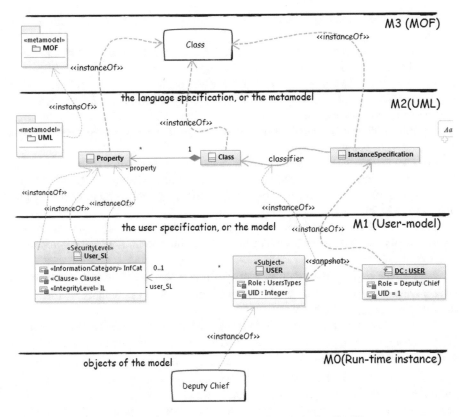

Fig. 1. UML four-layer meta-model hierarchy [8, 13]

2.4 An Approach to Properties Analysis of MLS Security-Design Models

Inspired by the work [12] we proposed our own language (meta)model MlsML and achieved formalization of restrictions for the specific BLP and Biba models. Our proposal relates mainly to possible formalization of restrictions in the OCL language, towards the relationship between the users (subjects), facilities, privileges, performed actions and certain states of the system. Contrary to the work [9], which also proposes its own tool that implements proposed methodology, we have based our solution on a typical CASE environment that has appropriate support for the UML models validation process (actually DSL) in the "starting" defined OCL limitations and UML models (expanded by the semantic action language). Our approach has been validated in the IBM RSA environment with Simulation Toolkit and Extension for Deployment Planning (for hardware modeling).

The approach to properties analysis of MLS security-design models and their instances leads to the evaluation and simulation. The integration of security models with models of systems described in UML, and the topology model [16] enables the simulation, which allows to verify/test the security properties of the designed MLS system software or/and the security policy models at the stage of analysis and modeling.

A secure system can be defined as a system that supports a specified set of policies. In the MLS approach, we support multiple high level policies as BLP or Biba and policy of domains separation secure. The basic aspects of our security policies can be defined in terms of the following provisions: (1) data isolation; (2) a limited provision for access to other domain's information - only if it is in compliance with the BLP or/and Biba policy; (3) shared resources of the system must be cleansed between security domain context switches.

We assume that the system under consideration consists of some types of resources from the ordered set $R = \{R_1, R_2, R_3, R_4, R_5\}$, where R_1, R_2, R_3, R_4, R_5 denotes accordingly processors' cores, memory segments (banks), input/output channels, graphic cards and hard discs.

Let $Proc = \{1, .., n\}$ and $Mem = \{1, .., m\}$ denote, accordingly, the set of processors numbers and memory segments numbers in the system. Let each of processors consists of r cores.

Matrix $B = \left[b_{ij}\right]_{n \times m}$ describes an access of processors to the memory segments in such way that $b[i,j] = 1$ iff i-th processor has direct access to the memory segment with number j and $b[i,j] = 0$ otherwise.

In the system we have the set of images of virtual machines treated as objects $VMI = \{vmi_1, \ldots, vmi_k\}$. The function $SL : VM \to C \times 2^{IC}$ define security levels assigned to virtual machines i.e. each of $vmi_i \in VMI$ is assigned a pair (c_i, A_i'), where $A_i' \subseteq IC$. We assume that in the MLS system may be a set of concurrently running virtual machines $VM = \{vm_0, vm_1, \ldots, vm_\kappa\}$, which of them constitutes a separate security domain with assigned $SL(vm_i)$; vm_0 is the home operating system with the virtual monitor manager (VMM) application.

Let Policy(x) [17] will be a function that returns the set of memory segments numbers from which information can flow into the specified segment x and Contents(x) determines the data values stores in the specified memory segment(s), and Current Domain defines the relevant state of the current executing vm_i.

On the base of results presented in the work [17] we can state that a MLS system is separation secure if the following holds:

For all $vm_i \in$ VM, for any pair of states, s1 and s2, of the composite system
And for every memory segment, seg, of the virtual machine:
if

    Contents (Policy(seg)) **in** s1 = Contents (Policy(seg)) **in** s2
    Current_Partitions of seg **in** s1 = Current_Partitions of seg **in** s2 /\
    Contents seg **in** s1 = Contents seg **in** s2
then

    Contents seg **in** (top-step s1) = Contents seg **in** (top-step s2)

In order to ensure separability of security domains, VMM has to implement appropriate algorithm of hardware resources allocation. Further we limit our consideration only to the verification of the resource allocation algorithm.

A virtual machine vm_i might request resources which will be represented by a vector Req_i, a value $Req_i[j] = l$ for $j \in \{1, \ldots, |R|\}$ means that it is necessary to allocate for virtual machine vm_i l units of a resource type R_j.

A current core processors allocation we will describe by a matrix $CP = [cp_{ij}]_{n \times r}$ and $cp[i,j] = l$ in the case when core j of i-th physical processor was assigned to virtual machine with the number l i.e. (vm_l).

When s-th virtual machine is started, the procedure $Alloc(Req_s, i)$ of allocating cores of i-th physical processor is carried out if following conditions (2–3) of separation of virtual machines (MLS domains) are hold i.e.:

$$Run(vm_s) \underset{\phi}{\rightarrow} Alloc(Req_s, i)$$

$$\phi : (\exists i \in Proc |\{j : cp[i,j] = 0\}| \geq Req_s[1]) \tag{2}$$
$$\wedge [\exists j \epsilon \{1, \ldots, r\} cp[i,j] = l] \Rightarrow [SL(vm_l) = SL(vm_s)]$$

or

$$(\exists i \in Proc \; \forall j \in \{1, \ldots, r\} : cp[i,j] = 0) \wedge (Req_s[1] \leq r) \tag{3}$$

Similarly, conditions of separation due to memory allocation in physical memory segments might be formulated. However, it can be easily observed that if each of memory banks can be accessed directly only by one processor (by a channel) i.e. when $\forall j \in \{1, .., m\} \sum_{i=1}^{n} b[i,j] = 1$ then constrains of domains separation due to memory allocation are equivalent to (1).

Example 1. Let $R = \{Pcores, Mseg, Chan, Gcards, Hdisks\}$ and $Proc = \{1, 2\}$, $r = 4$, $Mem = \{1, 2\}$. Assuming each processor have assigned one memory bank we obtain $B = \begin{bmatrix} 1 & 0 \\ 0 & 1 \end{bmatrix}$. Let there will be three virtual machines $VM = \{vm_1, vm_2, vm_3\}$ and $\overline{Req_1} = [3, 1, 1, 1, 1]$, $\overline{Req_2} = [2, 1, 1, 1, 1]$, $\overline{Req_3} = [2, 1, 1, 1, 1]$. Further we consider the case when $SC(vm_1) = (confidential, \{\alpha, \beta, \gamma\})$, $SC(vm_2) = (confidential, \{\alpha, \gamma\})$, $SC(vm_3) = (restricted, \{\alpha, \gamma\})$. Initial allowable system resources R are depicted by a vector $Z_0 = [8, 2, 2, 2, 3]$. Thus if the first virtual machine vm_2 is started the procedure $Alloc(Req_2, i)$ for $i = 1$ is executed and two cores of the processor number 1 are allocated for vm_2. If the next started virtual machine will be vm_3 the condition (1) is not hold thus the procedure $Alloc(Req_3, i)$ for $i = 2$ will be executed (because of condition (3)).

3 The Profile MlsML

The MlsML profile was developed as an extension of previously proposed solution [7, 15] with additional possibility of the integrity verification on the base of Biba model.

The profile in the UML language is a set of stereotypes and labels that define the meaning of designed entities within the problem of the field, which is an extension of the UML language. Therefore, we could say that the developed MlsML profile extends the semantics of UML language on the area of security modeling used in the construction of the MLS type systems.

The stereotypes belong to a fixed profile and expand the existing elements of UML model by changing their semantics and graphical notation. Stereotype defines which class of the UML (meta-class) is combined with its characteristics and limitations, and how the stereotypical elements could be combined with other elements.

The creation of a profile precedes the detailed analysis of the areas for which it is developed [17]. The result of this analysis in step 1 is a glossary of terms of the field, which in our project consists of: *Subject, Object, Action Requested, Permission, Action, Security Level, Integrity level, Integrity, Clause* and *Information Category* (the last three are the properties of concepts). In the next step (Step 2) verbal definitions are transformed into the domain class model. In the final step (Step 3) the model of domain is analyzed, and then its elements are mapped to stereotypes, classes and properties (Fig. 1), which leads us to UML metamodel (M2 level).

3.1   MLS Metamodel

Metamodel of security type BLP for MlsML profile is shown in (Fig. 2). In this metamodel four stereotypes were defined, including: two classes that describe the *Security and Integrity Level* in the BLP (and Biba) model and two of their properties, i.e. *Clause* or *Integrity* and *Information Category*, and restrictions are presented in the Fig. 2.

3.2   Using the MlsML Profile

The profile application we begin from building the model of M1 level (Fig. 1). The constrains in OCL language for the dominance function (checkDom()) is presented below.

---

context User2VMI_RequestedAction::checkDom():Boolean
pre: (self.vmi.vMI_SecurityLevel -> notEmpty()) **and (self.**user.user_SecurityLevel -> notEmpty())
post: result = (**self.**user.user_SecurityLevel.SLInfCat -> includesAll(**self.**vmi.vMI_SecurityLevel.SCInfCat)) **and**
(**if self.**user.user_SecurityLevel.SLClause >= vmi.vMI_SecurityLevel.SCClause **then** true **else** false **endif**)

It was assumed that in the implementation of the above constraints the attribute *Clause* from the set $\{Unclasified, Restricted, Confidential, Secret\}$ takes (accordingly) integer values from the set $\{0, 1, 2, 3\}$. The same profile may be used for the integrity verification, but in the current implementation (because of specific of SWSA project).

As can be seen in the Fig. 3 the MlsML profile enables creation of MLS lattice models and further their analysis for study the effects of the constraints of the applied policies. These policies can be formulated separately for BLP and Biba models, but also the combined use of the two models is permitted.

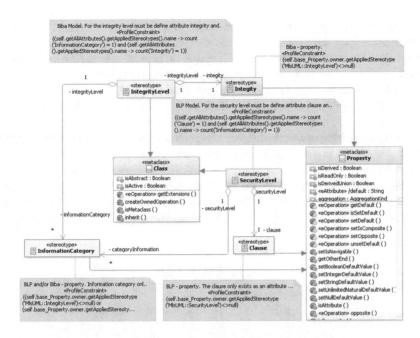

Fig. 2. Security profile (MAC – BLP & Biba models)

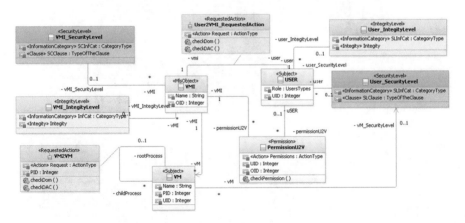

Fig. 3. M1 level for MlsML metamodel (BLP, Biba access control model)

3.3 Hardware Description with Topologies Model

Resources of SWSA which were used in the Example 1 (see Sect. 2.4) have been described as elements of topology, suitable for modeling hardware. These elements have been defined in the same way as the construction of the UML profile (see Sect. 3). However, to adequately describe this area it is better to use specialized DSL language for topology models [16].

The proposed approach to analyze resources of SWSA as the elements of topology, together with their description, has been developed as a plugin, which is attached to IBM RSA environment. Switching this plugin on enables new palette with topology elements which are an important component of the constructed simulator. Complete set of available topology units of SWSA are presented in the Fig. 4.

Allocation decisions, as referred to the Example 1, were mapped in the topology, by binding software components to hardware units: the VM2: one ProcessorUnit, two CoreUnit(s), one MemoryBankUnit, one GraphicCardUnit and one DiskMemoryUnit, and for other VMs in the same way according to defined resources requirements.

The result of the hardware design is SWSA topology model, which can be examined for compatibility with the defined rules (e.g. *At least one Core must be hosted on ProcessorUnit, Number of cores must be the power of 2*, etc.). The allocation of resources given in the example is verified using simulation, as described in Sect. 4.

a)

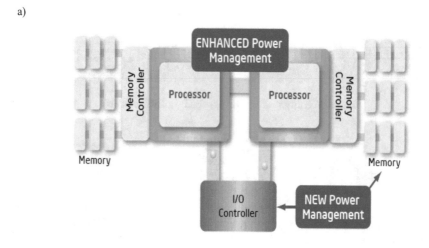

b)

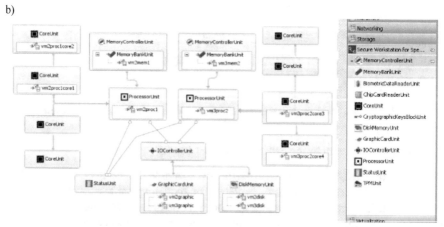

Fig. 4. Block diagram of the system based on the Xeon processor 5600 series and its representation in the topology model

4 Case Study: A Security Policy Model Verification and Simulation

In the following part of work, we will present an example to illustrate the use of the proposed method for the construction of the MLS security policy used in the SWSA project. The domination relationship is shown in the model as the operation of the User2VMI association class (Fig. 3) and in OCL language it was implemented as constraints that this operation must fulfill. We define constraints for a created model of security policy to run virtual machines by checking the domination relationship.

The main difficulty of the proposed approach to the verification of compliance with the rules of the MLS by Virtual Machine Monitor (VMM, see Sect. 4.1) was the development of a simulator. The simulator should allow not only to verify the compliance security policy, but also should make it possible to validate the correctness of the design of VMM (i.e. separation of resources, see Sect. 2).

Simulator is based on the capabilities of IBM Rational Software Architect 8.5: Simulation Toolkit 8.5 [18–20] and Extension for Deployment Planning [16]. Both of these capabilities enabled the authors to make the MlsML language profile as well as the palette of topology elements for SWSA which create the *framework* of the simulator. Based on this framework a simulator that can examine algorithms of VMM monitor, can be built.

The construction process begins with creating hardware model of SWSA, which we present in the topology diagram. In such diagram not only units and relations are included, but also its correctness rules (checked online, live). The construction of simulator ends with the creation of software model (components of SWSA), and its binding with the topology model (hardware) - manually or using drag and drop method.

An example of using the simulator, as a reference to Example 1 is as follows: *Create an input to the simulation process* (as that defined in Sects. 4.2 and 2.4); *Develop fragments of VMM behavior model* (which is the subject of the simulation); *Run simulation session*; *Select VMM scenario*; *Control the course of the simulation and collect its history* (Fig. 7); *Analyze the history of changes in status of resources, software components, and the state-word of the simulation - SSS*; *Develop report* (the output form simulation should clearly confirm of compliance with the security policies and separability (or lack thereof).

4.1 Secure Workstation for Special Application (SWSA) Project

In SWSA project to ensure the security of multilevel classified data an approach to develop software on the base of the virtualization technology for the separation of independent security domains was taken. To develop this type of MLS system the integration of available virtualization technology (software and hardware), application of formal methods for both ensuring and control of the confidentiality and integrity of data are needed. A natural way to build such systems is component approach, which assumes the use of ready and available hardware components and software, in particular virtualization packages (hypervisors) available as open source like Xen.

Developed within SWSA project software should allow for the simultaneous launch of several specific instances of operating systems on one PC (such as a workstation or server) designed to process data of different classification levels (e.g., public and proprietary), or to process the data in different systems, for which there need for separation of data.

The key element of SWSA is Virtual Machines Monitor (VMM), which is responsible for control of running virtual machines in accordance with defined MLS security policy and their switching to ensure the separation of resources. It was assumed that the proposed VMM software should make it possible to simultaneously launch several (of many possible) instances of special versions of operating systems on a single computer with the provision of: access control, separation of resources, cryptographic protection, and strict control of data flow.

In the SWSA environment one can distinguish three types of actors: the system administrator, the security officer and user. The security officer with the administrator and others are developing special security requirements of the system (the MLS security policy), and safe operation procedures.

4.2   SWSA Security Policy Verification – An Example

Example 2. The security policy specification for SWSA.

$Subjects(Users) = \{DC, CSO, OD\}$, where DC, CSO, OD denotes the SWSA users, accordingly, Deputy Chief, Commanding System Officer and Operation Director.

$Objects(Virtual machines) = \{VMI1, VMI2, VMI3\}$ where $VMIi, i \in \{1 \cdots 3\}$ denotes images of virtual machines VMi.

Let the set of numbers $InfCat = \{1\ldots6\}$ represent set of information category as follows 1 – Infrastructure, 2 – Strategy and Defense, 3 – Personnel, 4 – International Security, 5 – Commanding Systems, 6 – New Weapons.

In the system SWSA the ordered set C of clauses is used (from the lowest to highest):

$C = \{Unclassified(Uncl), Restricted(Res), Confidential(Conf), Secret(Sec)\}$.

The MLS security policy claims security level $SL(s) = (c, Infc)$ assigned to each of subject $s \in Subjects$ and security level $SL(o) = (c', Infc')$ assigned to each of object $o \in Objects$, where $c, c' \in C$ and $Infc, Infc' \subseteq InfCat$.

$$SL() \begin{bmatrix} DC \\ OD \\ CSO \end{bmatrix} = \begin{bmatrix} Sec & \{1..6\} \\ Sec & \{3,4,5,6\} \\ Sec & \{2,5,6\} \end{bmatrix}, \text{ and } SL() \begin{bmatrix} VMI1 \\ VMI2 \\ VMI3 \end{bmatrix} = \begin{bmatrix} Sec & \{1,2,3\} \\ Conf & \{3,4\} \\ Sec & \{5,6\} \end{bmatrix}$$

Assume that one of the actions from the set $Actions = \{Run, Modify\}$ may be requested by user, which may concern the selected virtual machine and users have assigned permissions to virtual machines from the set $Permissions = \{Read(R), ReadAttr(RA), Write(W)\}$. Let $Run(u, v)$ be a predicate that user u is allowed to take the action Run on a virtual machine v. Thus the following rules are holding $Run(u, v) \Rightarrow Perm(u, v) - \{R, RA\} \neq \emptyset$,

$Modify(u, v) \Rightarrow Perm(u, v) - \{R, RA, W\} \neq \emptyset$, where $Perm(u, v)$ is the subset of permissions assigned the user u to the virtual machine v.

Assuming that all users have granted permissions R, RA to all *VirtualMachines* we easily verify MLS security policy given in the example e.g. we prove that $Run(DC, VMI1) = true$, $Run(OD, VMI2) = false$ etc.

The presented model of M0 level (Fig. 5) is an instance of decision of security officer in the security policy matter, described in the Example 2.

The constructed policy is analyzed up to date (in the process of its construction, i.e., each insertion of a new item it is preceded by the checks of constraints) and subjected to a live and batch verification, enabling the correction of committed errors resulting from non-compliance of the pre-defined rules (defined by metamodel and model levels). Therefore, we can assume that as a result of the construction of the security policy with the use of the built environment we can create only correct policy. This is particularly important in the case of very complex systems with large number of subjects and objects. The Security Policy in critical systems (due to the protection of information) should be verified prior to the implementation of the system (Fig. 5).

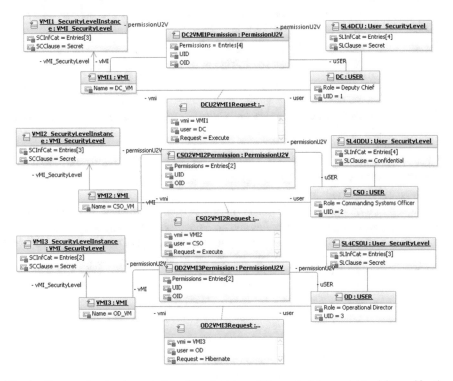

Fig. 5. Case study – implementation of the decision of Example 1 in special tool for verification MlsModel.

4.3   Analysis on Separability on the Base of an Example Scenario

The process of running of the virtual machine (Fig. 6) on the base of the virtual machine image $vmi_j, j\varepsilon\{1, 2, \ldots, K\}$ with defined security context $sc(vmi_j)$, by the subject (user) u_i with defined security level sl_i we will describe as $u_i \overset{RUN}{\rightarrow} vmi_j$. The process running of the virtual machine (subject) will be generated on the basis of the image vm_{j_k}, and its current security level (csl) for a subject vm_{j_k} will be defined as a pair $(clause, infCat)$ from the following dependency:

$$csl \sim = \langle \min(Clause(sl(u_i)), Clause(sc(vmi_j))), InfCat(sl(u_i)) \cap InfCat(sc(vmi_j)) \rangle,$$

(4)

where $InfCat(e)$ is the set of information categories of the element $e \in U \cup VMI$.

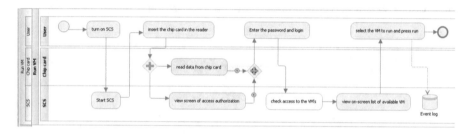

Fig. 6. The business model of a process of virtual machine running in the SWSA system

For verification of the models behavior we propose the use of the simulation mechanisms of UML models with the semantics action language as an extension. This is particularly important because the OCL language does not allow us to express constrains based on the states of models (there can be no changes in the characteristics of class instances (objects)). The diagram in Fig. 7 presents results of simulation of resources allocation as in the Example 1. These results (i.e. historic messages between VMs[2]) confirm VMM's behavior, which refuse to allocated requested resources to VM3 because of security violation (see step 12).

---

[2] The environment used in the work enables you to collect the simulation results in the following forms: history of messages sent between objects, traces of messages passing control flow, history of console records. It should be noted that capabilities of this environment may be extended with the use of UAL language.

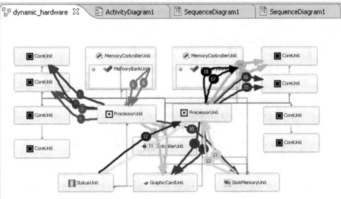

Fig. 7. The states verification of the process of running virtual machines by the model simulation

5   Summary

In the work we extend the MlsML profile in such a way that it enable modeling and verification of confidentiality and integrity of MLS system together. It is possible to investigate the real effects of the use of different Biba policies and combine them with BLP policies. We proposed and tested the complete environment based on the IBM RSA tool for MLS security policies verification, which can be easily used by security officers. The usefulness of this approach was confirmed in the completed SWSA project [21].

We demonstrated that the topology models may be successfully integrated with security-design MLS models for express and verification of the behavior of developed MLS system components by means of simulation in the deployment environment early i.e. in the modeling phase.

In this contribution we extend the method for multilevel security analysis to a Universal Model Language approach. Resorting to this means it allow to check of confidentiality and integrity of Multi level Security. The work is useful as it represents an import step forward to apply an assessed technique to any abstract system presented in the UML format. The proposed technique may be also applied in more general case as design of infrastructure for multiple level of security in service provisioning based on cloud computing.

References

1. Bell, D.E., La Padula, L.J.: Secure computer system: unified exposition and multics interpretation. ESD-TR-75-306. ESD/AFSC, Hanscom AFB, Bedford, MA (1976). http://csrc.nist.gov/publications/history/bell76.pdf. Accessed 24 June 2012
2. Bell, D.E.: Looking back at the Bell-La Padula model, Reston, VA (2005)
3. Biba, K.J.: Integrity consideration for secure computer system. Report MTR-3153 (1975)
4. Clark, D., Wilson, D.R.: A comparison of commercial and military computer security policies. In: Proceedings of the IEEE Symposium on Research in Security and Privacy, pp. 184–194 (1987)
5. Sandhu, R.S.: Lattice-based access control models. Computer **26**, 9–19 (1993)
6. Mouratidis, H., Giorgini, P., Manson, G.: When security meets software engineering: a case of modeling secure information systems. Inf. Syst. **30**(2005), 609–629 (2005)
7. Zieliński, Z., Stasiak, A., Dąbrowski, W.: A model driven method for multilevel security systems design. Przegląd Elektrotechniczny (Electr. Rev.) **2**, 120–125 (2012)
8. Basin, D., Clavel, M., Doser, J., Loddersted, T.: Model driven security: from UML models to access control infrastructures, vol. 15, no. 1, pp. 39–91 (2006)
9. Basin, D., Clavel, M., Doser, J., Egea, M.: Automated analysis of security-design models. Inf. Softw. Technol. **51**, 815–831 (2009)
10. Ahn, G.J., Shin, M.E.: Role-based authorization constraints specification using object constraint language. In: Proceedings of the 10th IEEE International Workshops on Enabling Technologies, WETICE 2001: IEEE Computer Society, Washington, DC, USA (2001)
11. Sohr, K., Ahn, G.J., Gogolla, M., Migge, L.: Specification and validation of authorization constraints using UML and OCL. In: Proceedings of the 10th European Symposium on Research in Computer Security (ESORICS 2005). Lecture Notes in Computer Science, vol. 3679, Springer (2005)
12. Jürjens, J.: UMLsec: extending UML for secure systems development. In: Jézéquel, J.-M., Hussmann, H., Cook, S. (eds.) UML 2002—The Unified Modeling Language. Lecture Notes in Computer Science, vol. 2460. Springer (2002)
13. Frankel, D.S.: Model Driven Architecture: Applying MDA to Enterprise Computing. Wiley, Hoboken (2003)
14. Zieliński, Z., Furtak, J., Chudzikiewicz, J., Stasiak, A., Brudka, M.: Secured workstation to process the data of different classification levels. J. Telecommun. Inf. Technol. **3**(2012), 5–12 (2012)
15. Stasiak, A., Zieliński, Z.: An approach to automated verification of multi-level security system models. In: Janusz, K. (ed.) Advances in Intelligent and Soft Computing. Springer (2013). ISSN: 1867-5662
16. Narinder, M.: Anatomy of a Topology Model Used in IBM Rational Software Architect Version 7.5, Part 2: Advanced Concepts. IBM, Armonk (2008)

17. Alves-Foss, J., Taylor, C., Paul Oman, P.: Multi-layered approach to security in high assurance systems. In: Proceedings of the 37th Hawaii International Conference on System Sciences—2004. IEEE (2004)

18. Mohlin, M.: Model Simulation in Rational Software Architect: Simulating UML Models. IBM, Armonk (2010)

19. Mohlin, M.: Model Simulation in Rational Software Architect: Communicating Models. IBM, Armonk (2010)

20. Anders, E.: Model Simulation in Rational Software Architect: Activity Simulation. IBM, Armonk (2010)

21. Kozakiewicz, A., Felkner, A., Zieliński, Z., Furtak, J., Brudka, M., Małowidzki, M.: Secure Workstation for Special Applications. Communications in Computer and Information Science, vol. 187, pp. 174–181. Springer, Berlin (2011)

Towards Definition of a Unified Domain Meta-model

Bogumiła Hnatkowska[(✉)] and Anita Walkowiak-Gall

Faculty of Computer Science and Management,
Wrocław University of Science and Technology,
Wyb. Wyspiańskiego 27, 50-370 Wrocław, Poland
bogumila.hnatkowska@pwr.edu.pl

Abstract. The domain model is one the most important artifacts produced at early stages of software development which can be reused efficiently during software design and implementation. Unfortunately, there is no clear explanation what should be included in the domain model to minimize modeling effort and maximize the potential later benefits. Authors of the paper try to address this gap by proposing a definition of a unified domain meta-model, which is the result of literature overview and authors' own experiences. The purpose of the literature analysis was to assess the existing definitions of domain models and to identify their common parts. On that basis, a unified meta-model was proposed. Its application in real projects confirmed the meta-model usefulness.

Keywords: Domain model · Concept model · Domain vocabulary
Meta-model

1 Introduction

Models play a crucial role in software engineering. They can be a "complete or partial description of systems" [18]. If the models present an enterprise, they are called business models. Business models can be built for different reasons, e.g. for business re-engineering, business creation, or business automation. They enable to [21]:

- understand the structure and dynamics of the organization,
- identify and understand current problems the organization has, as well as the areas of possible improvements,
- derive requirements for the future system supporting the organization.

Business modeling process, dependently on the context and needs, can have different scope. It can be limited to produce a domain model only, can aim in presentation of organization structure and its processes (processes' map), or can deliverer the full knowledge about the static and dynamic aspect of the modelled organization, including models presenting its actual state ("as-is" models) and/or its future state ("to be" models). Regardless of the chosen scope of business modeling, the domain model is the artifact that is always produced.

Domain model (its refined version) can be effectively reused during software development. Its importance is underlined in some approaches to software

© Springer Nature Switzerland AG 2019
P. Kosiuczenko and Z. Zieliński (Eds.): KKIO 2018, AISC 830, pp. 86–100, 2019.
https://doi.org/10.1007/978-3-319-99617-2_6

development, e.g. Domain-Driven Design [10] or Model Driven Development [19], and is reflected in the definition of design patterns, e.g. domain model pattern [11, p. 116]. The domain model can also be an input to database design activities [11].

Unfortunately, there is no clear explanation what should be included in a domain model at early stages of software development to maximize the later benefits with minimal effort at this phase. The paper will try to answer this question by literature overview and authors' experiences.

The rest of the paper is structured as followed. Section 2 presents the results of literature overview together with research questions and literature qualification criteria. Section 3 contains a proposed definition of the unified domain meta-model. Section 4 concludes the paper and proposes directions for further works.

2 What Is a Domain Model – Literature Overview

The notion of 'domain model' is not very often used in the literature in the meaning defined in the previous chapter where domain model is a part of business model created, for example, to support software design and implementation. One of the sources, in which the notion is used in such context, is Rational Unified Process (RUP) [13, 21]. RUP is an iterative, evolutionary methodology, which covers all basic stages of software development – from requirements analysis till testing. It also directly refers to a business modeling stage. Other positions are [1, 2] which describe, among others, the data modeling standards for products developed for the Natural Resource Sector (NRS) in Canada.

In some sources instead of 'domain model' the notion of 'concept model' (e.g. [16]) or 'structured business vocabulary' (e.g. [4, 17]) is introduced. Taking into account the purpose of these models, and when they are built in the case of business processes automation – all the notions: 'domain model', 'concept model', 'structured business vocabulary' can be treated as synonyms.

The above-mentioned sources were selected in accordance with the following assumptions:

- The source represents a methodology/standard/notation that covers at least business modeling stage and/or defines one of: 'domain model', 'concept model', 'structured business vocabulary'.
- The source is practicable, i.e. there is a community of methodology/standard/notation practitioners.

The book [13], according to Google Scholar, was cited more than 5000 times. SBVR is an OMG standard which is publicly available for ten years – it is a subject of plenty of scientific research papers, e.g. [9] has almost 100 citations. The similar statement can be made about Ross's books [16, 17].

2.1 Domain Model – RUP Definition

In [21], the domain model is defined as an "incomplete business object model" focusing on business entities, business deliverables, important business events and

relationships among them. Such model lacks for example information about responsibilities of business workers (roles).

To understand this definition one must define business object model first. The
business object model is an "object model describing the realization of business use-
cases. It serves as an abstraction of how business workers and business entities need to
be related and how they need to collaborate in order to perform the business" [21]. In
further we limit our consideration to the static aspect of business use-case realization,
which is represented by a UML class diagram [3].

"The domain model typically shows the major business entities, their functional
responsibilities, and the relationships among the entities. It also provides a high level
description of the data that each entity provides" [15]. Operations are presented at a
general level, without parameters. However, in some works that refer to RUP, e.g. [20]
no operations are mentioned, and attributes are used only optionally.

The elements that can be visualized by a domain model according to [8, p. 131],
are:

- attributes with types (both elements are optional); it is allowed to define a type
  attribute if no more than one object needs to have a direct access to it, otherwise a
  relationship between classes should be used [21] (e.g. *Person* class can have
  Address attribute if it is the only place where *Address* is used)
- different type of association including aggregation and composition
- multiplicities,
- names of associations and roles (option),
- a direction of association arrow (option)

The alternative to defining a domain model is to write a business vocabulary [8,
21]. Business entities, their features, and relationships among them are candidate
elements for the system model [8, 21] including definition of system use-cases [8].

2.2 Domain Model – NRS Data Modeling Standards

"The domain data model, also known as a conceptual data model, is a high-level
representation of business information within a project and/or application" [2]. It is an
optional artifact which purpose is to "help guide and communicate data requirements"
[1] amongst various business stakeholders.

"The domain model consists of a class diagram and related descriptions. The data
elements on the model will typically describe a person, place, thing, event or concept
for which the NRS business has an interest" [2]. The diagram is technology agnostic
and concentrates on major classes that describe business data requirements held within
NRS, as well as data shared with or by external agencies. Classes' symbols can be filled
with colors (e.g. white for business class) to enhance the diagram readability. The
previous version of the standard [1] also had defined some guidelines about the diagram content, e.g.:

- important/obvious attributes may be presented, however without type or multiplicity prescription;

- each attribute, as a class it is contained in, must have a "clear, concise and understandable textual definition of the purpose of the attribute"; "business keys showing the unique business identifier for the classes may be added where known";
- associations between classes should have multiplicities defined, however, their names or role names are optional (role names are used as if they were association names); aggregations and compositions are not permitted.

Domain model is an input for data design performed later at logical and physical level.

2.3   Concept Model/Structured Business Vocabulary – R. Ross Definition

A structured business vocabulary is "the set of terms and their definitions, along with all wordings, that organizes operational business know-how. (…) The system of meanings that the words in a structured business vocabulary represent is called a concept model" [17]. A concept model refers to the static aspect of information describing an organization (a problem domain), and/or information that the organization is going to manage [16].

Business vocabulary defines concepts (the main information units) which "give structure to basic business knowledge" [16]. There are two types of concepts: *noun concepts*, designated by terms, and *verb concepts/fact types* designated by verbs or verb phrases.

A concept and its particular meaning should be precisely defined in a business-oriented form, and be present in the vocabulary. The definition should take into account synonyms and multiple languages, to support recognition of distinct contexts of usage and communities with different speech preferences.

A noun concept is a noun or a noun phrase that is recognized and used in business communications, which may have specific meaning within the context of the organization domain. It should refer to:

- something fundamental to business, which cannot be derived or computed from any other terms,
- a 'thing' whose instances are discrete, i.e. can be counted, and
- "a thing we can know something about, rather than how something happens" [16].

The noun concept refers to the class (type) rather than to class' instances.

A verb concept defines the connection between noun concepts. The part that a noun concept plays in the context of the connection is a *role*. A verb concept is expressed using verbs or verb phrases relating to appropriate terms (designating noun concepts or their roles). The noun-and-verb construction – a phrase of a "predictable type that permits sentences, especially expressing business rules, to be made for business operations" [16] – is called wording.

Connection involves two noun concepts in general. However, there is possible to define a verb concept involving more than two noun concepts, i.e. an n-ary fact type, or a verb concept concerning only a single noun concept, i.e. a unary fact type. A unary fact type, synonymous with a property, denotes a quality or trait of a noun concept.

Similarly to a noun concept, a verb concept refers to a type of connections rather than an individual connection. There are several important types of connection between concepts, i.e. predefined sentences:

- property association (a binary verb concept in which one thing is closely tied to the meaning or understanding of another): (thing) *has* (thing),
- categorization: (class of thing) *is a category of* (class of thing),
- composition: (whole) *consists of* (parts),
- assortment: (thing) *is an instance of* (class of thing).

To present a concept model in a graphical form, R. Ross proposes the ConceptSpeak[TM] notation.

A concept model defines the structured businesses vocabulary around the project (in particular IT project) and should act as a communication tool for everyone involved. Therefore, a model can be seen as a ubiquitous language, accessible and understandable by everyone who is involved with the project, crucial for expressing business requirements effectively [16]. It should be mentioned, that ubiquitous language is an important part of Domain-Driven Design [10], it should be directly derived from the concept model.

2.4 Business Vocabulary – SBVR Definition

SBVR [4] is an OMG standard that formalizes the definition of business vocabularies and business rules. It separates expressions used to communicate (e.g. a sequence of characters, diagrams) from their meanings. These two are connected by representations ("each representation ties one expression to one meaning" [4]).

The vocabulary is defined as a "set of designations (such as terms and names) and verb concept wordings primarily drawn from a single language to express concepts within a body of shared meanings" [4].

The SBVR document can be perceived as a formal version of RuleSpeak (Ross).[1]

Concepts are "units of knowledge created by a unique combination of characteristics" [4]. There are two main types of concepts: noun concepts (are meanings of nouns or noun phrases), and verb concepts (are meanings of verb phrases that involve one or more verb concept roles).

Noun concepts are classified into [4]:

- general concepts (designated by terms, e.g. car, which serve for classification of things on the basis of their common properties),
- roles (designated by terms, e.g. rental car, here: abstraction of a car playing a part in an instance of a verb concept, e.g. branch stores rental car), and
- individual concepts (instances of general concepts, designated by names, e.g. Los Angeles).

---

[1] www.BRSolutions.com.

Noun concepts can be related by categorization relationship (generalization/specialization). Constraints, representing the fact that categories are mutually exclusive or that they cover the whole domain can be documented separately in vocabulary.

General concept can be a concept type if it serves for specialization of another concept. A specific concept type is categorization type "whose instances are always categories of a given concept", e.g. 'organization function' (categorization type) serves for specialization of the 'organization unit' concept. Instances of 'organization function' categorization type include, e.g. national HQ, branch, local area [6].

Verb concepts are classified into:

- binary verb concepts (involves exactly two roles),
- unary verb concepts (characteristics),
- individual verb concepts (all roles refer to individual noun concepts),
- general verb concepts (has at least one role not being individual noun concept).

Two noun concepts can be related with a binary verb concept, e.g. site is in country. The relationship represents a simple unconstrained fact (i.e., 'many-to-many' and 'optional' in both directions) [7]. The more restrictive rules that apply to these constructs can be expressed as business rules (out of scope).

There is also possible to define n-ary verb concept involving more than two roles or to define a unary verb concept called characteristics (e.g. car is small).

Verb concepts can be objectified by a given general concept (at most one) [4].

To sum up, there are following verb concept types:

- Association and property association – a verb concept that has more than one role used to express connections between things; in the case of property association the verb concept is used to describe properties of things, e.g. person has eye color
- Characteristics – a verb concept with exactly one role, e.g. car movement is round-trip
- Partitive verb concept (part-whole verb concept) – a verb concept to represent a part-whole relationship between things.

Besides concepts, the meaning of things can be expressed with propositions, which enable to define e.g. classifications (the instance of a given individual noun concept is an instance of a given general concept), categorizations, or characterizations. Categorization is a statement expressing the fact "that a given general concept specializes a given general concept" [4]. The general concept can be divided into categories by the categorization scheme, e.g. the concept "person" can be categorized by gender into "male" and "female". The type whose instances are categories is called categorization type (concept type in general). It can be partial or complete (if it covers all possibilities). A specific type of categorization scheme is segmentation (partitioning) "whose contained categories are complete and disjoint" [4].

There exists a graphical representation of business vocabularies, defined in [7]. It is also mapping to the UML (Annex C [4]).

Business models and SBVR are pointed out as an important input for MDA approach, which can be translated to Platform Independent Model (PIM) [5]. Especially, the application of SBVR standard is important for "designers of information systems that support business vocabularies or automate business rules" [6].

2.5   Summary

This section compares the notions described in the preceding sections. The results are shown in Table 1. To make the comparison possible, authors decided to relate particular elements to a normalized domain model notation. This notation assumes that domain model is represented by a graph, which consists of two types of basic elements: nodes and edges (at specification level), and node instances and edge instances (at instance level).

Nodes are the primary building blocks of the model, presenting basic concepts from the domain of interest. Each node must be somehow defined (e.g. with a unique name, and/or some additional textual documentation). A node can have some properties assigned which, on their own, can have their definitions (unique names, textual documentation). A node can have responsibilities (functions) assigned with their definitions. If a notation recommends usage of a specialized node it is also mentioned in the table.

Edges serve for representing connections among nodes. The difference with a typical graph is that the edge can connect more than two nodes. Similarly to nodes, they should be somehow defined. The nature of allowed connections is described in edge characteristics section. The edge can have properties on its own. It has two or more ends documented in three subsequent rows: 'edge end', 'edge end definition', and 'edge end characteristics'. The specific type of edge is generalization relationship which connects a more general and a more specific concept.

Rows filled with the gray show the common elements in all notations (dark gray – the full match, light gray – the partial match).

Different names are used to describe the same notion, i.e. "domain model", "conceptual model" [8, 21], "domain data model", "conceptual data model" [1, 2], "concept model", "structured business vocabulary" [16, 17], "business vocabulary" [4]. Each of the considered approaches recommends one or more graphical notations to describe the basic entities and relationships among them, most (three from four) use the UML for that purpose.

There is an agreement what to model at a general level. Concepts are represented by classes or their equivalents (terms). Their semantics is covered by a textual description. Concept's properties are modeled either by attributes (RUP, NRS) or by unary verb concepts and/or by a special type of binary verb concepts 'has' i.e. property associations (Ross, SBVR). Types of properties are not obligatory (in NRS they are prohibited). Only major properties are to be modeled if necessary, and their role, e.g. business key, may be mentioned, see NRS. Only one notation (NRS) recommends specification of an entity with a set of predefined values (also known as reference data or an enumerated list, equivalent of an enumeration type in the UML).

Concept responsibilities are never modeled with one exception, i.e. RUP.

Relationships between concepts play a crucial role and are always represented. The relationship's name is always to be defined, however, in NRS the names of association ends are used for that purpose. Three notations allow defining two relationship's names for the same relationship, i.e. NRS, Ross, SBVR, the first with the use of role names (e.g. "An Employee works for 1 Department", "A Department employs 0 to many Employees" [1], underlined elements are relationship names), the other two with a

Table 1. Comparison of domain model content recommended by considered notations.

	RUP	NRS Data Model-ing Standard	Ross	SBVR
Model name	Domain model/ Conceptual model	Domain data model/ Conceptual data model	Concept model/ Structured busi-ness vocabulary	Business vocabu-lary
Notation	UML class dia-gram + related de-scriptions	UML class dia-gram + related de-scriptions	ConceptSpeakTM	SBVR SE/Concept Dia-gram Graphic No-tation/Mapping to UML
Elements at specification level				
Node	Class/Business Entity	Class	Noun concept	Noun concept
Node defini-tion	Class name Textual descrip-tion	Class name Textual descrip-tion	Term Definition of the noun concept	Term Definition of the noun concept
Node property	Attribute	Attribute	1. Unary fact type/ Property 2. Binary fact type 'has'/ property as-sociation	1. Unary verb con-cept 2. Property associ-ation defined with "of", "has"
Node property defini-tion	Attribute name	Attribute name	1. Verbs/ Verb phrases 2. Term	1. Verb concept wording, e.g. car is small 2. Term
	Attribute type	Standard suffix where the attribute is defined as a spe-cific type, e.g. Date for date at-tributes	1. Unary fact type is always Bool-ean-valued 2. NA	1. Unary verb con-cept is represented as Boolean value in MOF and UML 2. Concept type
	Textual descrip-tion	Textual descrip-tion Business key Color	1. Definition of the property	2. Definition of the noun concept
Node re-sponsi-bilities	Operation		NA	NA
Node re-sponsi-bility defini-tion	Operation Parame-ters		NA	NA
Special nodes	NA	List of values	NA	NA
Edge	Association	Association	Verb concept rep-resented by a wording / Fact type	Verb concept rep-resented by a wording/ Associa-tion

(continued)

Table 1. (*continued*)

Edge definition	Association name	NA (indirectly defined by role names)	1. Verbs/ Verb phrases for binary verb concept 2. Sentences illustrating n-ary fact type	Verb used as a name
	Association direction arrow	NA (indirectly defined by role names)	Direction arrow for binary verb concept	Synonym form of verb concept wording for binary verb concept
	NA	NA	Definition of verb concept	Definition of verb concept
Edge characteristics	1.Binary (bi-directional) 2. n-ary 3. Aggregation/Composition	1. Binary (bi-directional)	1. Binary (bi-directional) 2. n-ary, 3. Composition/ fact type 'consists of'	1. Binary 2. n-ary 3. Part-whole verb concept
Edge properties	NA	NA	Verb concept objectification	Verb concept objectification
Edge end	Role	Role	Role	1. Role 2. Verb concept role (wording with has)
Edge end definition	Role name	Role name (represents the association name)	Term	Term
Edge end properties	Multiplicity	Multiplicity	Default multiplicity *	Default multiplicity *
Edge: Generalization relationship	Generalization	Generalization	Categorization/Fact type 'is a category of'	Categorization
Generalization characteristics	NA	NA	Group of categories/Categorization scheme	Categorization scheme Segmentation Concept type
Elements at instance level				
Node instance	NA	NA	Instance/Individual noun concept	Individual noun concept
Node instance definition	NA	NA	Name Definition of an instance Assortment / fact type 'is an instance of'	Name Definition of individual concept
Edge instance	NA	NA	Fact	Fact

synonymous form of a verb phrase. A binary relationship is the primary association type, present in all considered notations. Only NRS does not support n-ary and composition/aggregation relationships.

Relationship's ends are typically modeled by roles represented by nouns (NRS is the exception, in which roles are verbs). Multiplicity is defined either explicitly (RUP, NRS) or implicitly (Ross, SBVR). In the latter case, the default multiplicity is many, and more restricted demands are defined by business rules. Relationship's properties (in UML they can be represented with the use of association class) are mentioned in two approaches, i.e. Ross, SBVR. Each notation uses generalizations to model "is a" relationship between concepts.

Instances of concepts are present in two notions, i.e. Ross, SBVR.

The domain model is perceived as an important input for software development. Names and textual descriptions of its elements form a valuable glossary (ubiquitous language shared among all involved stakeholders). The elements of the domain model are refined during the design phase where necessary properties and responsibilities are added, so, in consequence, the domain model is translated into a part of a software system. It is also true in the context of database modeling in which domain model is transformed into a logical data model. It is worth to mention that the domain model is also seen as a part of MDA approach in which formally expressed vocabulary with business rules can be automatically transformed into Platform-Independent Model.

3   Definition of a Unified Domain Meta-model

This section defines a unified domain meta-model (UDM hereafter). The meta-model was constructed with the use of below described method:

- The elements present in all considered notations (see Table 1, rows filled with gray) were included to UDM automatically. That was done in pre-assumption that common parts are a kind of consensus in domain modelling area.
- The elements present in any subset of notations were discussed against their usability and occurrence frequency in the commercial projects, the authors take part in; on the basis of that analysis the elements were either excluded or included to the UDM. The justification for decisions is also shown.

The UDM is expressed by UML class diagram notation – see Table 2, column Class Diagram. Class diagram notation is chosen because of its common usage in the development process. In result, it will support the integration with development artifacts such as business process models, business requirements and/or use cases.

Elements of the UDM are enlisted in Table 2. This version introduces elements at specification level only.

The entity type and the relationship type are primary specification level representatives. The entity type corresponds to the graph node which maps to a class or a noun concept in the considered notations – see Table 1, and is denoted by UML *Class* meta-class. It is defined by a unique entity type's name (the UML *name* meta-attribute), a description (the UML *ownedComment* meta-attribute), and a set of entity type's properties (the UML *ownedAttributes* meta-attribute). A property of entity type is

Table 2. Definition of unified domain meta-model.

Model name	Unified domain model	Class diagram
Notation	UML class diagram and related textual descriptions	
Elements at specification level		
Node	Entity type	Class
Node definition	Entity type name	NamedElement::name : String [0..1]
	Textual description	Element:: ownedComment : Comment [0..*]
Node property	Property	Property – StructuredClassifier:: ownedAttribute : Property [0..*]
Node property definition	Property name	NamedElement::name : String [0..1]
	Type	TypedElement::type : Type [0..1]
	Description	Element::ownedComment : Comment [0..*]
Special nodes	Enumerator type + a set of values	Enumeration, Enumeration::ownedLiteral : EnumerationLiteral [0..*]
Edge	Relationship type	Association, AssociationClass
Edge definition	Name of relationship type Direction	NamedElement::name : String [0..1]
	Description	Element::ownedComment:Comment [0..*]
Edge characteristics	Binary, n-ary,	memberEnd : Property [2..*]
Edge properties	Properties of entity type	Class in AssociationClass
Edge end	Property	Property – Association:: memberEnd : Property [2..*]
Edge end definition	Property name	NamedElement::name : String [0..1]
Edge end properties	Multiplicity	MultiplicityElement::/lower: Integer [1..1], MultiplicityElement::/upper: Unlimited-Natural [1..1]
Generalization relationship	Generalization	Generalization
Generalization characteristics	Generalization scheme	GeneralizationSet GeneralizationSet::isCovering : Boolean [1..1] = false GeneralizationSet::isDisjoint : Boolean [1..1] = false

defined by a unique name (the UML *name* meta-attribute), a textual description (the UML *ownedComment* meta-attribute) and optionally a type (UML *type* meta-attribute). The type is defined only if the property holds values of an enumerator type. Due to being an essential part of domain knowledge, the authors include enumerators in the UDM. The enumerator type corresponds to the special graph node which maps to a list of values (for NRS) – see Table 1, and is denoted by UML *Enumeration* meta-class.

The relationship type corresponds to the graph edge and maps to an association or a verb concept in the considered notations – see Table 1. It is denoted by UML *Association* meta-class. When a part of domain knowledge represented by the relationship requires additional information, the relationship type has properties to represent the information. In such case, it is mapped to UML *AssociationClass* meta-class.

The relationship type is defined by a relationship type's name (the UML *name* meta-attribute), a textual description (the UML *ownedComment* meta-attribute), a set of entity type's properties (the UML *ownedEnd* meta-attribute) – option, and multiplicities of the properties (the UML *lower* and *upper* meta-attributes).

As in the real-world, a relationship can have properties on its own (e.g. marriage date, and place or employment status or salary), the UDM recommends using UML association classes to model such a case. It seems to be a more natural solution than those suggested in NRS, where a separate class is used for that purpose.

Generalization is one of the most important relationships modeled at a business level. After Ross and SBVR we recommend defining for it a generalization scheme describing the nature of the generalization set, e.g. if it is covering or complete. By default, each generalization is interpreted as overlapping and incomplete.

4 UDM Instance Example

Figure 1 outlines an example of UDM prepared on the basis of [14] where the same model is presented in the ConceptSpeak^{TM} notation. The notation allows e.g. describing association's name in two directions. In UDM, the association's name can be specified in one direction only. However, the name in the opposite direction could be specified in the association description. On the other hand, UDM allows to define constraints for generalization set (e.g. {complete, disjoint}), and multiplicities for

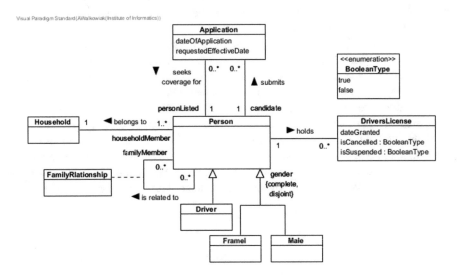

Fig. 1. UDM instance example.

association ends. Enumeration type for *Boolean* values was introduced to keep the UDM instance consistent with its source [14].

Due to the limited size of the paper, the example skips textual descriptions of nodes and relationships on the diagram.

5 Conclusions and Further Works

The paper faces the problem of domain model creation. It is built for many purposes, among others, domain model:

- forms the basis for other business modeling artifacts, e.g. business rules or business process,
- serves for identification and definition of all concepts the organization uses within its business processes,
- allows verification of consistent domain understanding among all stakeholders,
- enables effective communication between domain experts, business analysts, architects and software designers.

The paper tries to answer the question what should be contained in a domain model at early stages of the development of a software product to increase the model usefulness during software design stage. The answer is given as a result of literature overview combined with authors' experience, gained during business modeling stage performed for commercial IT projects. The analysis allowed to find out the elements recommended by all considered notations, however, the list was extended with some elements recommended only by some standards, e.g. application of enumeration types with a predefined list of literal values [1].

The main conclusion is that the domain model must be supported with textual descriptions (defining semantics). All basic concepts (represented by UML classes and properties) have to be formally defined to avoid misunderstandings. Synonymous forms for both noun concepts as well as verb concepts (relationships) are welcomed. In the future, they will enable clear definition of business rules, especially when a natural language other than English (e.g. Polish) with less predictable grammar is used.

The UDM is a good choice when structural features are to be modeled. For domains strongly dependent on time (e.g. real-time systems), it could be not enough.

In further we are going to experimentally validate the usefulness of UDM, to compare it with results of systematic literature overview (in progress), and on this basis to define the unified domain meta-model more formally which would enable its comparison to ontology expressiveness, e.g. OWL2, and automatic translation between different representations. Domain ontologies could be potentially treated as a valuable source for domain models [12]. The other considered direction is a systematic literature overview.

References

1. NRS Data Modeling Standards with EA, Version: 3.1.0, 29 November 2016
2. NRS Data Modeling Standards with EA, Version: 3.2.0, 28 March 2017. https://www2.gov. bc.ca/assets/gov/british-columbians-our-governments/services-policies-for-government/ information-technology/standards/natural-resource-sector/sdlc/stanards/nrs_data_modeling_ standards_ea.pdf
3. OMG Unified Modeling Language 2.5, 01 March 2015. http://www.omg.org/spec/UML/2.5/
4. OMG Semantics of Business Vocabulary and Business Rules (SBVR)^{TM}, v. 1.4 (2017). https://www.omg.org/spec/SBVR/1.4/
5. OMG Semantics of Business Vocabulary and Business Rules (SBVR), v. 1.4, Annex E—Overview of the Approach (2016). https://www.omg.org/spec/SBVR/1.4/Annex-E–Overview-of-the-Approach/PDF
6. OMG Semantics of Business Vocabulary and Business Rules (SBVR), v. 1.4, Annex G—EU-Rent Example (2016). https://www.omg.org/spec/SBVR/1.4/Annex-G–EU-Rent-Example/PDF
7. OMG Semantics of Business Vocabulary and Business Rules (SBVR), v. 1.4, Annex I—EU Concept Diagram Graphic Notation (2016). https://www.omg.org/spec/SBVR/1.4/Annex-I–Concept-Diagram-Graphic-Notation/PDF
8. Arlow, J.: UML and the Unified Process, Practical Object-Oriented Analysis & Design. Addison-Wesley, Boston (2002)
9. Bajwa, I.S., Lee, M.G., Bordbar, B.: SBVR business rules generation from natural language specification. In: AAAI Spring Symposium: AI for Business Agility (2011)
10. Evans, E.: Domain-Driven Design: Tackling Complexity in the Heart of Software. Addison-Wesley Professional, Boston (2004)
11. Fowler, M.: Patterns of Enterprise Application Architecture. Addison-Wesley Professional, Boston (2002)
12. Hnatkowska, B., Huzar, Z., Tuzinkiewicz, L., Dubielewicz, I.: A new ontology-based approach for construction of domain model. In: Intelligent Information and Database Systems: 9th Asian Conference. Lecture Notes in Computer Science. Lecture Notes in Artificial Intelligence, vol. 10191 (2017). https://doi.org/10.1007/978-3-319-54472-4_8. ISSN 0302-9743
13. Kruchten, P.: The Rational Unified-Process: An Introduction. Addison-Wessley Professional, Boston (2004)
14. Lam, G.: The top 10 mistakes business analysts make in capturing business rules. IPMA professional development events (2011). http://ipma-wa.com/prof_dev/2011/Gladys_Lam_Ten_Mistakes.pdf
15. Menard, R.: Domain modeling: leveraging the heart of RUP for straight through processing. IBM developerWorks (2003). https://www.ibm.com/developerworks/rational/library/2234. html
16. Ross, R.: Business Rule Concepts, Getting to the Point of Knowledge, 4th edn. Business Rule Solutions, LLC, Houston (2013)
17. Ross, R., Lam, G.: Structured business vocabulary: concept models. Bus. Rules J. **17**(11) (2016). http://www.brcommunity.com/a2016/b880.html
18. Rumbaugh, J., Jacobson, I., Booch, G.: The Unified Modeling Language Reference Manual, 2nd edn. Addison-Wesley, Boston (2005)

19. Stahl, T., Voelter, M., Czarnecki, K.: Model-Driven Software Development: Technology, Engineering, Management. Wiley, Hoboken (2006)
20. Suarez, E., Delgado, M., Vidal, E.: Transformation of a process business model to domain model. In: Proceedings of the World Congress on Engineering, 2–4 July, 2008, vol. 1, pp. 165–169. Newseood Ltd., International Association of Engineers, London (2008)
21. Rational Unified Process. http://sce.uhcl.edu/helm/rationalunifiedprocess

Hybrid Agile Method for Management of Software Creation

Agata Smoczyńska, Michał Pawlak, and Aneta Poniszewska-Marańda[✉]

Institute of Information Technology, Lodz University of Technology, Lodz, Poland
{agata.smoczynska,michal.pawlak}@edu.p.lodz.pl,
aneta.poniszewska-maranda@p.lodz.pl

Abstract. Agile approach to software development provided a good alternative for heavyweight traditional methods. However, it has its drawbacks, that are discussed more and more often, such as: inefficiency in large teams and projects, difficulty in accurate task estimation, lack of fully defined requirements at the beginning of the project, or being "overly extreme" with its principles and practices. The paper presents the new hybrid method in agile approach, which improves shortcomings of the chosen methodologies. Additionally, a tool for supporting project creation using the proposed method in agile approach is created.

Keywords: Agile methodologies · Agile software development
Scrum · Extreme programming

1 Introduction

According to the current situation where the software and market develop and change very quickly, the traditional software development models, like waterfall, are obsolete. Heavyweight methods of software development, that were often overly regulated, planned, and too strictly managed, were commonly used. However, they were too restrictive, which caused a need for more lightweight methods to appear [1].

These lightweight methods started appearing and evolving during the 1990s [1]. They encouraged discarding strict formalism of traditional methods, instead encouraging adaptability, using iterative-incremental approach, putting focus on the team work and more dynamic process of software creation, which allowed more reliable and flexible software creation process. Some of such lightweight methods were created before the Agile Manifesto was written in 2001; however, they are included in agile software development methods.

Agile software development has gained popularity and currently is widely recognized. As often as it is accepted, however, it is discarded as not suitable for "serious" companies or not effective for large and complex projects. Agile approach to software development provided a good alternative for heavyweight traditional methods. However, it has its drawbacks, that are discussed more and more often, such as: inefficiency in large teams and projects, difficulty in

© Springer Nature Switzerland AG 2019
P. Kosiuczenko and Z. Zieliński (Eds.): KKIO 2018, AISC 830, pp. 101–115, 2019.
https://doi.org/10.1007/978-3-319-99617-2_7

accurate task estimation, lack of fully defined requirements at the beginning of the project, or being "overly extreme" with its principles and practices.

Despite these flaws, agile approach to software development remains widely used. Adopting them brings several benefits, such as improvement of transparency of the project, cost and risk reduction or better control over the project costs. Agile methodology, that is the most suitable for an organization, can be chosen from multiple existing agile methodologies.

Scrum is currently the most widely used agile methodology. It provides a framework for project managements and focuses on business side of the project management and leadership. Thus, Scrum can be adapted to wide variety of project types. Extreme Programming, on the other hand, focuses mainly on software development. It focuses on providing good programming and engineering practices.

The paper presents the new hybrid method in agile approach, which improves shortcomings of the chosen methodologies. Additionally, a tool for supporting project creation using the proposed method in agile approach is created. Moreover, the proposed method is to be tested on a project with a small scope to verify its basic assumptions.

The paper is organised in the following way: Sect. 2 presents the general concepts of agile software development and project management with special regard to benefits and challenges of adopting the agile project management. Section 3 describes the proposed method in agile approach – the values, practices, process phases, team roles and responsibilities, meetings and documents are discussed. Section 4 deals with an overview of the prototype of support tool for the proposed method of agile approach.

2   Agile Software Development and Management

The concept of software lifecycle is the basis of the software development. The software lifecycle in general is a time period, beginning with an idea for software, creating it, and ending when it no longer is available. The software development lifecycle (SDLC) consists of several phases. Figure 1 shows the phases of software development lifecycle.

Agile, according to [3], is "[t]he ability to create and respond to change in order to succeed in an uncertain and turbulent environment". Agile Manifesto and the 12 principles confirm and provide deeper understanding for this statement. *Agile software development* is an "umbrella term" for several software development methodologies, methods and practices, which share a common vision and core values and use similar iterative-incremental approach [3–5]. The goal of this method of software development is quick production and delivery of high-quality products. It is crucial to acknowledge that the requirements will change often during the project and adapt to such changes accordingly [1,6].

It should be noted that agile methodologies are quite often lightweight, especially in comparison to traditional waterfall methods of software development. In fact, they were created as an alternative to the waterfall model [6]. Additionally, agile methodologies emphasise "empowering people to collaborate and

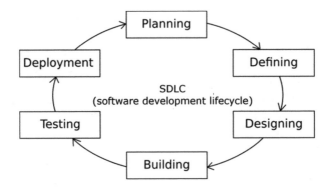

Fig. 1. Various phases of typical software development lifecycle, based on [2]

make decisions together quickly and effectively" [4]. Agile software development methods are suitable and recommended for small teams, which develop small- or medium-sized software, usually for business, in which the most crucial factor is the speed of delivery, as well as adaptation to frequent changes. Small size of the team is recommended to decrease or even prevent problems with communication between team members, thus preventing also need for overly extensive documentation [6].

2.1 Benefits of Adopting Agile Project Management

Introducing agile approach to a project management brings several advantages. They are following [6]: speed of obtained solutions, plasticity, risk management, cost reduction and better control over costs, high quality, right product, transparency.

Speed of obtained solutions is listed as the first advantage of adopting the agile approach. The agile approach allows to ensure quick, constant and efficient delivery of created software [1,6]. This is achieved thanks to splitting larger tasks to smaller ones, focusing on problems and tasks which are the most valuable to the client and setting shorter deadlines for manageable chunks of work.

Plasticity means that solutions and further improvements are made according to the client's expectations and requirements, preferably with close cooperation with the client. It is inscribed in the project's nature. *Risk management* includes implementing solutions as simply as possible, so that a client would be able to quickly use them and verify, whether proposed solutions meet their requirements and needs. So obtaining continuous feedback from the client and close cooperation with the client is encouraged [1,6].

Cost reduction and better control over costs are possible thanks to applying good agile practices, such as frequent contact with the client, swift response to changing requirements, and delivering further parts of working software in short iterations. *High quality* of delivered software is obtained by applying continuous evolution – frequent testing of the product [6].

Right product means that the delivered software meets all needs of the client and is the perfect solution to their problems. *Transparency* (or visibility) of the project requires close and frequent cooperation with the client. Showing them the changes introduced to the project, obtaining their feedback as soon as possible and discussing with them further possible changes in the project and requirements enable to achieve the clarity of everyone's expectations (in project terms) and reduces the risk of misunderstandings [7].

The annual state of agile survey, conducted by VersionOne [8], presents what respondents expected from adopting agile into their companies' development process and which of these expectations were met. According to respondents, adopting agile to company's development process helps to better manage changing requirements, improve project's transparency and increase speed of delivery. It has positive impact on the team's productivity and morale. Moreover, companies of vast majority of respondents (98%) achieved success using agile approach to project management.

2.2   Challenges of Adopting Agile Project Management

Similar to all existing solutions, agile project managements has its drawbacks. Despite successes of projects which source is adopting agile, mentioned in previous section, there are challenges that companies and teams experienced while adopting agile project management. The respondents of the state of agile survey [8] had chosen which challenges and drawbacks were met during introducing agile into company's development process.

One of the major challenges for adopting agile methodologies is experience needed for successfully adopting them in the company's development process. Fowler admits that some techniques may be hard to comprehend until they were attempted to be done multiple times, and strongly advices finding a good mentor and closely following their advice [11]. The development team and each of clients should be trained to know the basics of agile methods. Such trainings may be quite costly and may increase the overall project cost [9–11].

Ineffective collaboration can have the most damaging impact on project. If the client is not interested in collaboration, achieving full advantages of adopting agile becomes impossible; poor client participation directly affects quality of the project and may lead even to its failure [9,11]. Fowler warns that imposing agile methods on team that is unwilling to do so is going to be a serious struggle and should never be done [11].

Close cooperation with the client may not be possible for ever team, especially for distributed (or remote) teams. It is required that all project participants can attend meetings [10]. It may be impossible, especially for distributed teams located in different countries or even in different parts of the world [12,13]. Since it is not possible to gather all project participants in one meeting, due to distance or time zone differences, workarounds have to be established, such as meetings in certain time slots [13]. Additionally there is risk of frequent meetings disrupting the "flow" of programmers [12].

2.3   Hybrid Method of Scrum-XP

Both *Scrum* and *Extreme Programming* share similarities, such as core values, team roles and responsibilities, phases or scope. Both methodologies work well for small or medium teams (preferably no larger than 10 developers) [14]. They are said to work well together – they are quite similar, often adopt each other's practices, and complement each other well [15]. However, there are significant differences between Scrum and Extreme Programming, which allow them to be used simultaneously and successfully in the project.

Scrum provides framework for creating, delivering and maintaining complex products and projects. It can be easily adapted to any type of project, for it does not focus on any specific area, for example software engineering. It rather focuses on business side of project management and leadership, as well as ensuring transparency of the project. No programming practices are introduced with this methodology.

Extreme Programming, on the other hand, provides detailed practices and suggestions regarding good programming practices, as well as for designing a developed system and testing it. It "aims to produce higher quality software, and higher quality of life for the development team" [3]. Thus XP focuses on providing the best engineering practices while neglecting management practices.

This major difference – with XP focusing on programming practices, and Scrum focusing rather on management practices – is what allows these two methodologies to be successfully used together in a single project. Combined, they provide the client and the development team with a structure that allows creating software that best meets the client's needs, and allows achieving highest possible quality in each iteration while taking advantage of business opportunity.

3   Proposition of Scrum-XP Hybrid Method in Agile Approach

According to a survey conducted by VersionOne [8], *hybrid methods of Scrum and XP* is one of the most commonly used approaches in respondents' organizations (Fig. 2). However, it is not listed amongst agile methodologies, thus it can be said that the hybrid method of Scrum-XP is not considered as a methodology. Thus, there are no "official" guidelines or rules how to use such hybrid in a project; this is left for organization to decide. Thus, for example, two IT companies, which both use hybrid method of Scrum-XP, may in fact use two different methodologies for managing their projects.

The aim of this paper is to present a new formalized agile approach, under name **Scrum-XP Hybrid (SXP Hybrid)**, combining strengths and good practices of both *Scrum* and *Extreme Programming* to cancel out or reduce their weaknesses. Besides providing a set of improved good practices, SXP Hybrid proposes team roles and responsibilities or process phases, based on those existing in Scrum and XP, as well as proposes a set of documents useful for the project. SXP Hybrid focuses exclusively on software creation projects. Thus, it may not

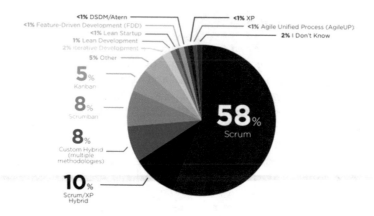

Fig. 2. Agile methodologies used in organizations according to the survey [8]

be suitable for projects from different fields. Additionally, it is intended for small teams and small scope of the project.

3.1 Values

SXP Hybrid combines core values of Scrum and Extreme Programming (Fig. 3):

1. Face-to-face communication is the most reliable form of communication.
2. Feedback is crucial for the good project and allows to introduce necessary changes. Openness allows all concerns to be addressed freely.
3. Respect is deserved by everyone. Each team member is equally valuable.
4. Courage is necessary to tell the truth about problems, estimates and failures. There is no need to fear, because no one works alone.
5. Simplicity and putting focus only on one thing at a time is the key to successful project, as is maximizing work not done.

Additionally, following three "pillars" are recommended: transparency, inspection and adaptation. Openness (core value of Scrum) is crucial for obtaining a good feedback (core value of XP), thus both values are shown on the same level and connected with a straight line.

3.2 Practices

SXP Hybrid focuses exclusively on software-related projects. Thus, good programming practices provided by Extreme Programming are incorporated into it. Figure 4 presents the described practices divided into two categories: coding practices and team practices. Coding practices are practices that focus on improving programming and easier code maintenance. Team practices focus on ensuring a good working environment and conditions for the team, as well as maintaining a good atmosphere within the team.

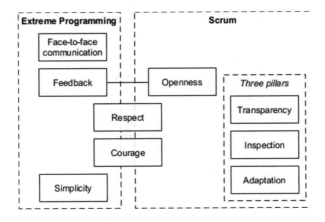

Fig. 3. Core values of *SXP Hybrid* approach

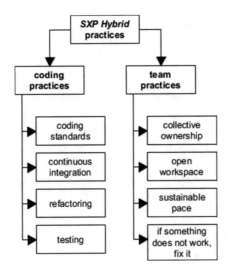

Fig. 4. Practices proposed by *SXP Hybrid* approach

3.3 Process

SXP Hybrid proposes five phases: exploration and design, planning, development, testing and deployment. These phases utilize elements of phases of Scrum and XP as well as the six steps of agile project management (Fig. 5). Exploration and design phase needs to occur only once, at the beginning of the project. Next four phases create a complete cycle that is repeated with each next iteration.

During the ***exploration and design phase***, features and requirements (functional and non-functional) are gathered, technology and tools are chosen, and initial system design is prepared. At the beginning of this phase, a kick-off meeting is held. The client, the developers, the team leader and the product

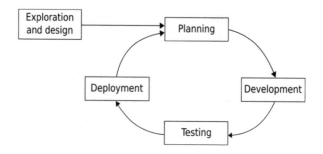

Fig. 5. Five phases of *SXP Hybrid* approach

owner should participate. They discuss the project in greater detail to provide the development team with basic necessary information.

As a result of exploration and design phase, both the client and the development team ought to: share a common vision on the project, know what the expected result is, and have an idea how the finished product will be implemented. Five documents should be created:

- *vision document*, providing a short and concise overview on project: why is there need for this product, what are intended features of the product, etc.,
- *project specification document*,
- *product backlog*, containing the prioritized list of the client's requirements,
- *initial version of a product plan*, providing an overview on more technical aspects of the project,
- *coding standards document*, covering the practices and coding standards suitable for chosen programming language and technology.

During the **planning phase**, plans for the current iteration are made. The *iteration planning meeting* is held. During it the iteration plan for the current iteration is created and items for the iteration backlog are selected. Each of the user stories selected for the iteration should be discussed to clear out any ambiguities.

As a result of the planning phase, both the client and the developers ought to: know which functionalities are most important for the client, and what are the tasks corresponding to them. A single document should be created:

- *iteration plan document*, containing the iteration backlog, providing task cards, task estimates, as well as target release dates.

During the **development phase**, the work on the product is done in iterative cycles. During each iteration the developers work on implementing user stories from iteration backlog for this iteration. Iterations ought to have constant length, to make planning and measuring progress simple and reliable. Recommended duration is 1–3 weeks. Iteration must end after a specified period of time, even if not all tasks are done.

Every day, a daily stand-up meeting is conducted. If possible, the client should be present during it. Alternatively, in case of iterations longer than one week, a weekly walk-through meeting could be held at the end of the week.

As a result of the iteration (the development phase), a working software ready to be tested should be obtained. Optionally, a set of necessary documents can be created, for example detailing how certain aspect of project work, or which database table stores which data.

The **testing phase** occurs concurrently to the development phase. The project is constantly tested using automated unit tests, as well as manually by the developers. However, after the end of iteration the obtained software should be tested more thoroughly.

During this phase the whole product is tested, manually and using integration tests, to find and report any shortcomings to be fixed as soon as possible. After these minor errors are fixed, acceptance tests created by the client are run. If acceptance tests are passed, the system is ready to be deployed. Software testing is a very complicated process. It is not independent – it is limited by and conditioned by the development process.

During the **deployment phase**, the work on the increment is concluded and preparations for the next iterations are started to being made. After work on the increment is concluded, an *iteration review meeting* should be held. The software must be shown to the client, even if it has not passed the acceptance test – obtaining the client's feedback is crucial for further development.

During this phase the *iteration retrospective meeting* is held. It allows the team to discuss successes and failures of the iterations, or possible improvements. The deployment phase completes a cycle of the project management. The feedback and information gathered during this phase are a base for the next cycle of the project management, starting with the planning meeting for the next iteration.

3.4   Team Roles and Responsibilities

Team roles and responsibilities proposed in *SXP Hybrid* approach are based on team roles present in Scrum and XP – five roles are proposed: client, product owner, team leader, developer and tracker. Figure 6 presents an overview of them, detailing their dependencies and relationships, and shortly explaining responsibilities of each role.

3.5   Meetings

The meetings proposed by *SXP Hybrid* ought to eliminate the need of adding more and more meetings, thus potentially eliminating the Scrum flaw: adding more and more unnecessary meetings.

The **kick-off meeting** is held at the beginning of the project, during the exploration and design phase. The goal of this meeting is to determine expected effect and establish a common view of the project. Each project participant

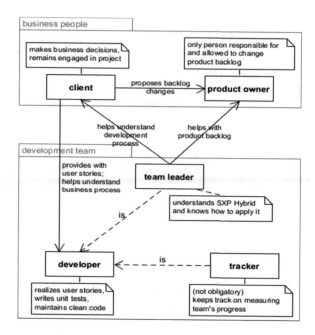

Fig. 6. Dependencies between team roles proposed by *SXP Hybrid* approach

ought to be present during this meeting. The client presents their expectations: what software they want the team to create, what they want the software to do, and what are their initial requirements.

The ***iteration planning meeting*** occurs at the beginning of each iteration, during the planning phase. The goal of this meeting is to prepare an iteration plan and iteration backlog: select user stories from the product backlog to be implemented in the new iteration, determine who will be responsible for each task, and during the planning game estimate effort and time required to complete the task.

The ***daily stand-up*** is a short, daily event, occurring during each iteration. The main goal of this meeting is to facilitate communication between all team members. Each team member reports what they done yesterday, what they plan to do today and what obstacles and difficulties they met.

The ***iteration review meeting*** occurs during the deployment phase. The goal of this meeting is to show the client what was achieved during this iteration, gather feedback from the client, and make adjustments to the product plan, product backlog and other documents if necessary. The client, the product owner, and the development team should participate.

The ***iteration retrospective meeting*** occurs during the deployment phase, after the end of the iteration review meeting. The goal of this meeting is to provide the development team with the opportunity to analyse their strengths and weaknesses and create a plan for improvements. Only the development team should be present on this meeting.

3.6   Documents

Although Extreme Programming encourages maximizing the work not done, the documentation it proposes (tests, user story and task cards, iteration and release plans) is too minimalistic. Thus, *SXP Hybrid* proposes including documentation inspired by Scrum documentation (namely, Sprint and Product Backlogs), as well as documents not mentioned in both methodologies. The documents proposed by *SXP Hybrid* are as follows:

- vision document,
- project specification document,
- coding standards document,
- product plan document,
- product backlog,
- iteration plan document.

Project overview, its basic assumptions and requirements, or design decisions should be known to all project participants: the client, the product owner and the developers. Thus, a ***vision document*** is proposed. Its main goal is to give all project participants clear and common vision of project. The document should be kept in place which all project participants have access to. If any project participant wants a quick reminder about the project goals, those information are easily available to them.

The ***project specification document*** serves two goals: to provide a detailed overview on product functionalities and technical aspects. It is divided into two parts: functional specification and technical specification. The *functional specification* part describes all functionalities available in the product. The descriptions should be as detailed as possible, so the programmer at the first glance knows what should be achieved and how. The *technical specification* part describes all technical aspects of the product: used technologies, environments, tools, or architecture.

Coding standards is a programming practice encouraged by Extreme Programming. The ***coding standards document*** is a helping, obligatory document. Intention behind it is to keep in one place coding standards for all programming languages used in the project. The document should be kept in place which all developers has access to. Every programmer should familiarize themselves with this document and apply gained knowledge during coding.

The ***product plan document*** is the equivalent of XP's release plan document. As such, it serves similar purpose and fulfils similar goal. It contains information about the product that was not suitable for the vision and project specification documents, such as the team organization, or project overview and scope. This document keeps information about all past iterations; they are presented in a table with their objectives, start and end dates, and target velocity.

The ***product backlog*** contains an ordered list of user stories (functionalities and requirements to be implemented) written by the client, which can be modified and extended. This document is the basis for the project. It evolves together with the project. The only person responsible for and allowed to edit this document is the product owner.

The aim of *iteration plan document* is to clearly state the goal of the current iteration: what functionalities the client wants to be implemented, and what tasks were assigned to them. The iteration plan contains a list of prioritized user stories that were selected for this iteration. It should contain detailed overview of tasks: their names and descriptions, assignees, priorities, estimates, state, and release dates.

SXP Hybrid is a method of agile approach intended for small teams and project scopes, combining strengths and good practices of Scrum and Extreme Programming to cancel out or lessen their weaknesses. However, it can be also used for medium or large projects dividing such projects into smaller sub-projects managed by SXP Hybrid, the same as it is realised using Scrum or XP methods. SXP Hybrid proposes programming practices, team roles and responsibilities, meetings and a set of documents for developing a software product.

4   SXP Hybrid Support Tool

There exist many tools for managing software development process, such as Jira, Microsoft Team Foundation Server or Zoho Projects, that may be more suitable for one company than to other. SXP Hybrid acknowledges this fact and does not rely on any specific management tool.

However, trying to simplify the process by introducing certain amount of automatization may have several benefits. This may be especially helpful for novices, who are not certain how to approach some aspects of project management. More experienced people also could benefit from it. For example, a tool which helps with creating documentation would speed up the process and reduce time needed for this task. A *SXP Hybrid support tool* was created with that in mind.

The prototype of the SXP Hybrid support tool provides the activities defined in SXP Hybrid approach described in the previous section. One of these activities is the function of creating the user story and managing the product backlog. They are presented in more detail in Fig. 7. The "Add user story to product backlog" use case is one of the the most complex. As can be seen in Fig. 7, it requires that user story is written, priority for it is selected, and estimated effort is calculated (which requires providing three estimation values: optimistic estimate, most likely estimate and pessimistic estimate). These things can be done in any order. However, if any of them is missing, a user story cannot be added to the product backlog.

To *add a new user story* to the product backlog, the user has to make a several steps. All steps must be performed or the user story will not be created and added. Firstly, the user story must be written. The user has to fill three text boxes, located within the "User story" panel, with suggested parts of the user

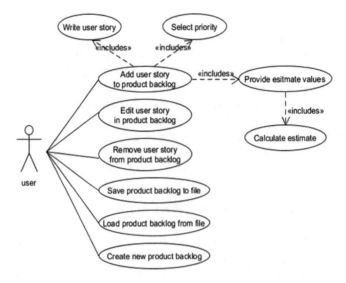

Fig. 7. Use cases for "Add user story to product backlog" function of the SXP Hybrid support tool

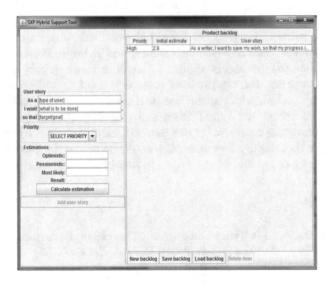

Fig. 8. "Add user story to product backlog" function in the SXP Hybrid support tool

story. The user input is combined with the fixed parts, so the result can look as the following:

> *As a writer, I want to save my work, so that my progress is not lost.*

Then the priority must be selected from the drop-down list. Three choices are possible: low, medium and high. Estimated effort must be calculated. The user has to provide optimistic, pessimistic and most likely efforts in their respective text boxes ("Estimations" panel). If all values were correct, the calculated estimate value appears next to the "Result:". Finally, the "Add user story" button has to be clicked. The user story is added to the product backlog, and all fields in user story generator form are cleared out (Fig. 8).

The SXP Hybrid support tool is a tool dedicated exclusively to SXP Hybrid method of agile approach. It means that the tool truly provides a support for creating software using this method, and any changes made to SXP Hybrid will be reflected in it. The tool itself is "light", simple and portable. It operates on files and does not rely on any complicated technology or framework.

5   Conclusions and Future Works

The paper presented the hybrid method in agile approach, which combines strengths and good practices of both Scrum and Extreme Programming to cancel out or lessen their weaknesses, as well as addresses some of critique of both methodologies. Moreover, *SXP Hybrid* suggests team roles and responsibilities, as well as project phases, documentation and meetings.

SXP Hybrid, however, was created with software development projects in mind. As such, it focuses exclusively on software and programming issues. Thus it may not be suitable for projects from different fields. Additionally, it is recommended that the team of developers should not be larger than about 10 individuals, and that the method is applied to small or medium projects. However, it is possible to divide the big projects to several smaller "subprojects" – each of the "subprojects" may be realized by such small teams using *SXP Hybrid*.

SXP Hybrid worked for a small team and a small project scope (programme for private, non-commercial use for one person). However, *SXP Hybrid* ought to be tested more thoroughly. The next step would be to use the proposed method for bigger project scope and bigger team, preferably in industrial environment.

References

1. Martin, R.C.: Agile Software Development, Principles, Patterns, and Practices. Pearson Higher Education; International Edition, Harlow (2013)
2. TutorialsPoint: SDLC – Overview. https://www.tutorialspoint.com/sdlc/sdlc_overview.htm. Accessed 5 Dec 2017
3. Agile Alliance: What is Agile Software Development? https://www.agilealliance.org/agile101/. Accessed 9 July 2017
4. VersionOne: What is Agile? Learn About Agile Software Development. https://www.versionone.com/agile-101/. Accessed 19 Apr 2017
5. Collier, K.W.: What is a self-organizing team? In: Collier, K. (ed.) Agile Analytics: A Value-Driven Approach to Business Intelligence and Data Warehousing, 1st edn. Addison-Wesley Professional, Upper Saddle River (2011)
6. Stellman Andrew, A., Greene, J.: Learning Agile: Understanding Scrum, XP, Lean, and Kanban. O'Reilly Media, Sebastopol (2013)
7. Rasmusson, J.: The Agile Samurai. The Pragmatic Bookshelf, Dallas (2010)

8. VersionOne: 11th Annual State of Agile Report, 6 April 2017. https://explore. versionone.com/state-of-agile/versionone-11th-annual-state-of-agile-report-2. Accessed 20 Apr 2017

9. Adell, L.: Benefits & Pitfalls of using Scrum software development methodology, 11 April 2013. http://www.belatrixsf.com/blog/benefits-pitfalls-of-using-scrum-software-development-methodology/. Accessed 7 Dec 2017

10. Haunts, S.: Advantages and Disadvantages of Agile Software Development, 14 December 2014. https://stephenhaunts.com/2014/12/19/advantages-and-disadvantages-of-agile-software-development/. Accessed 7 Dec 2017

11. Fowler, M.: The New Methodology, 13 December 2005. https://martinfowler.com/ articles/newMethodology.html. Accessed 7 Dec 2017

12. Gray, A.: A Criticism of Scrum, 24 October 2015. https://www.aaron-gray.com/ a-criticism-of-scrum/. Accessed 7 Dec 2017

13. Jammalamadaka, K., Ramakrishna, V.: Agile software development and challenges. IJRET Int. J. Res. Eng. Technol. **2**(8), 125–129 (2013)

14. Schwaber, K.: Agile Project Management with Scrum. Microsoft Press, Redmond (2004)

15. Cohn, M.: Scrum & XP: Better Together, April 2014. https://www.scrumalliance. org/community/spotlight/mike-cohn/april-2014/scrum-xp-better-together. Accessed 8 Dec 2017

System Monitoring and Performance

The Performance Analysis of Web Applications Based on Virtual DOM and Reactive User Interfaces

Dariusz Chęć[1] and Ziemowit Nowak[2(✉)]

[1] NeuroSYS Sp. z o. o., ul. Rybacka 7, 53-656 Wrocław, Poland
d.chec@neurosys.com
[2] Faculty of Computer Science and Management,
Wrocław University of Science and Technology,
Wybrzeże Wyspiańskiego 27, 50-370 Wrocław, Poland
ziemowit.nowak@pwr.edu.pl

Abstract. This paper contains the analysis of modern web applications. These applications are Single Page Applications (SPA) based on virtual DOM (VDOM) and reactive user interfaces. In order to perform a performance analysis for these complex web applications, two architectures were suggested. The architectures are capable of capturing asynchronous event streams processed within the data flow between the interface and the data model. The data flow is controlled by the available operators, principles and models of the reactive programming paradigm. The research shows that loading applications using virtual DOM requires relatively longer time associated with the VDOM structure construction process, while processing operations on the previously loaded user interface structure is much simpler and more efficient than the native methods. In addition, the functional reactive programming in JavaScript is conducive to the development of scalable web applications.

Keywords: Virtual DOM · Functional reactive programming · React.js
Cycle.js

1 Introduction

Since the time the decision was first made to mark Web sites as Web 2.0, the development of hardware, network and software technologies have been constantly observed. Web 2.0 concepts which characterized websites from the beginning of the current century began to be referred to as web applications. The turning point in this process was the popularization of Asynchronous JavaScript And XML (AJAX) – the new model of request handling for a web server based on asynchronous queries. Soon AJAX implementations began to appear in the first popular JavaScript libraries such as jQuery, Prototype, and MooTools. Those libraries offered a set of auxiliary functions facilitating the processing of the so-called Document Object Model (DOM). In a short time, the community hailed them as the guarantor of good practices – they were the first to provide a certain conventional pattern in the process of designing and implementing user interfaces. Over the years, the phenomenon of the disappearance of

© Springer Nature Switzerland AG 2019
P. Kosiuczenko and Z. Zieliński (Eds.): KKIO 2018, AISC 830, pp. 119–134, 2019.
https://doi.org/10.1007/978-3-319-99617-2_8

the traditional layered architecture of web applications in favour of Single Page Application (SPA) was observed [8, 10, 14]. The decision was made to transfer some of the business logic to be processed on the client's side, thus freeing the computing power of WWW servers, e.g. Apache, Microsoft ISS, nginx, which significantly improved their performance.

The increase in the complexity of business logic on the client's side and the need to process it, are further issues that the community had to face. Providing reliable and efficient Internet solutions serving millions of users a day involved questioning the long-term assumptions regarding the construction of web applications. On 8th March 2014, during a conference organized by Facebook Inc., software engineer J. Chen presented a unidirectional architecture of data flow between the model and the application presentation layer. The presented approach was the result of abandoning the traditional Model-View-Controller (MVC) paradigm in favour of user interfaces built using components. Application divisions for state and stateless components (presentation) controlled by one instance of data, i.e. application status, increased control over actions caused by users. Application scalability increased, and so did – as a result – the integrity of data presented by many interfaces, on many different devices. The author of the lecture presented the problem of synchronization of messages exchanged in Messenger as a case study of the web application and the native mobile application in the Android system [13].

Ensuring the consistency and speed of exchanged data between different types of interfaces turned out to be a performance challenge for web browsers, especially their mobile versions. In 2013, a team of Facebook Inc. engineers presented the first version of the React.js library. The library's task was to handle a large number of operations performed on the DOM tree in short intervals. The concept of operation was understood as frequent calling methods of the Element object specified in the specification called DOM (Core) Level 1 issued by the W3C Consortium [1]. These methods are responsible for inserting, replacing or deleting a specified node, i.e. an HTML tag to the list, or an attribute of such a tag. The list is stored in the form of a tree structure. It was decided to represent a list of elements in the form of JavaScript object literal. It was noted that the current recursive list processing operations can be performed on the basis of the function of comparing two objects where the first object represents the input state of the operation call and the other the state following its completion. The result of the function directly points to the elements that should be reflown and then repainted by the web browser. This type of structure is known as the Virtual Document Object Model (VDOM) [4, 6]. To date, many implementations have been created of the presented mechanism in the form of libraries or frameworks providing complete solutions in this field.

Modern web applications are less and less reminiscent of monolithic architectures, but are rather their opposites. Dispersed web systems operate in an asynchronous way, using the Representational State Transfer (REST) architecture in external and internal system communication (e.g. microservices). They collect data from many external sources, including images and multimedia provided via the Content Delivery Network (CDN), thus they are interrogated by many external websites, e.g. Google Analytics, API of social media. Proven reactive programming patterns are used to support so many asynchronous data.

The user interfaces created in this application model must handle the flow of asynchronous data in a special way. They should, among others, respond to events coming from a web server, read data shared by local databases, and monitor user events (actions performed using input-output devices). To work with data flowing in an asynchronous way, a multiplatform collection of libraries called ReactiveX is available in the form of a free software license. The first implementation of the library was developed for the .NET platform and C# programming language in 2011. By using the RxJS extension – the version of the JavaScript library it is possible to capture any kind of asynchronous events on the client's side (the web browser) and react to their occurrences in the expected manner. Creating reactive user interfaces using the API provided by RxJS is related to the use of patterns:

- observer pattern,
- iterator pattern,
- functional programming paradigm.

Reactivity understood in the context of creating user interfaces in the JavaScript language is referred to as Functional and Reactive Programming (FRP) [5].

2 Motivation

Each year, cyclical conferences are organized around the world by the largest companies in the industry, e.g. Facebook, Google, Netflix, Airbnb, during which the community presents problems that their websites are struggling with. These websites usually offer high-quality services to millions of users per day, and the Internet technologies they use become publicly-shared community projects. Web browsers, programming libraries, frameworks, or development tools become a marketing showcase promoted by those giants in generally available services for open repositories. The multitude of solutions in this area causes difficulties in choosing the right technology and application architecture. The previously mentioned React.js library, being the most popular VDOM implementation, is the second most popular repository on GitHub.com.

The elements of the JavaScript language extension proposed by the ReactiveX library on 10th January 2017 appeared in Stage 1 Draft as a proposal for the official language standard [7]. The process consists of five stages and is supervised by the TC39 unit of the Ecma International organization responsible for the development and maintenance of standards for libraries expanding language capabilities. In the near future, reactive tools for processing asynchronous data streams can become a native part of the language. Like the *Promises* pattern introduced in the sixth version of the ECMAScript specification, June 2015.

Web applications run on the client's side release the web server from the need to generate HTML code and thus implement business logic. As a result, the timing of the process of displaying and processing operations on the user interface becomes problematic due to the need to capture all asynchronous events as well as tags describing specific processes occurring in the browser window.

Bearing in mind the above-mentioned issues, the authors set themselves the following goal:

- an overview of libraries and tools (frameworks) used to build reactive user interfaces,
- suggesting a web application architecture capable of representing the virtual object structure of a document model (VDOM) in contrast to the traditional model,
- examining the efficiency of the prepared solution in opposition to classic interfaces built using the native DOM JavaScript API built into browsers.

3 Measurement of SPA Performance

3.1 Performance Timeline

The life cycle of modern SPA internet applications strongly depends on asynchronous events. Virtually each time a user calls an action, the browser will execute the Java-Script code in response, re-rendering and painting of the updated parts of the webpage will occur. Navigation Timing specification recommendations [15] defines a series of events (metrics) which describe the order of operations in chronological order. The calls of these events are recorded by the browser once, until the browser finishes its task and triggers the *loadEventEnd* event. Subsequent operations recording the values of those metrics will occur when the page is reloaded, which significantly hampers the examination of the efficiency of SPA-type internet applications.

Another problem of performance testing is the situation in which the source code of the application's business logic starts just after the moment the application displays the so-called Critical Rendering Path, and the browser triggers the *loadEventEnd* event again. In this situation, performance measurement is not reliable, because most of the business functions of the application will request the data provided by the web server layer (back-end) API in an asynchronous way. In such a situation Navigation Timing will not register any values for events and metrics, because the *loadEventEnd* event was executed earlier.

To enable the performance measurement of modern web applications, the Web Performance working group developed in December 2013 the Performance Timeline document [9]. This document neither defines metrics nor provides detailed character-istics for application events that occur during the life cycle. The aim of its creation was to standardize the interface for downloading metrics through web browsers and tools created to collect this type of metrics. On the basis of this document, tools for profiling the processing efficiency of client-side applications provided by web browsers were created. One of them is Google Chrome Timeline, which was later renamed Chrome DevTools Performance [3]. Due to the friendly interface, the researcher can easily display events in chronological order, filter them according to the time of their exe-cution and analyse the tasks divided into groups with the ability to view any level of complexity of a given operation in a very precise way. The most important event groups, and consequently the group metrics, are as follows:

- **Loading** – events responsible for the conversion and analysis (parsing) of HTML code to determine the object structure of the document model (DOM),
- **Scripting** – events recording the execution of JavaScript source code along with the browser's allocation control and memory release operations,
- **Rendering** – events evoked to process the page layout and to develop a CSSOM and DOM render tree,
- **Painting** – events responsible for rasterizing the result page to a bitmap format and displaying it in a web browser.

The individual components of the group metrics are listed in Table 1.

Table 1. Components of group metrics provided by the performance tools

Metrics	Group	Description of the measured interval
ParseStyleSheet	Loading	Processing of CSS code and generation of the CSSOM model
ParseHTML	Loading	Conversion, tokenization, lexical analysis and creation of the DOM model
EvaluateScript	Scripting	Interpretation of JavaScript code
Event	Scripting	Listening and executing JavaScript code events
FunctionCall	Scripting	Execution of JavaScript code
Layout	Rendering	Creating the page layout (creating the rendering tree based on the DOM tree)
RecalculateStyle	Rendering	Reprocessing (refreshing) the CSS code and generating the CSSOM model
UpdateLayerTree	Rendering	Preparing the layout for HTML nodes, determining their position, flowing relative to other elements
UpdateLayoutTree	Rendering	Preparing the page layout, determining the position of elements in the process of creating a render tree based on the DOM
Composite Layers	Painting	Merging layers (subtree of the rendering tree) over the GPU and rendering pixels on the screen
Paint	Painting	Rasterization of the displayed page to a bitmap

Other popular Internet browsers: Firefox, Internet Explorer, Edge, Safari, Opera are equipped with the same functions for performance measurement with a difference in the construction of interfaces. Coherence in the way of capturing individual events is dictated by the Performance Timeline recommendation standardized by the Web performance working group.

3.2 Tools Supporting Automation of Measurements

The creators of web browsers have developed dedicated versions of protocols thanks to which it is possible to download all registered measurements performed by the web browser and the application running within the current tab. The collected metrics are available to the programmer in the format provided by the protocol's API and can be freely processed by the tools generating reports. Due to the discrepancies and lack of standardization in the way of capturing performance-related metrics, the community came up with *remotedebug.org* initiative, whose task is to integrate individual specifications:

- Google Chrome Debugger Protocol,
- WebKit Remote Debug Protocol,
- Mozilla Debugging Protocol.

The RemoteDebug initiative was put forward by the originator K. Auchenberg at the international *Full Frontal* session in 2013, during which he gave a lecture titled "Our Web Development Workflow is Completely Broken" [2]. To date, no universal tool for automated measurements that would unify the common version of the metrics captured for each web browser has been developed.

Browser-perf. It is a Command Line Interface (CLI) application that runs in a multi-platform Node.js environment. The task of the application is to communicate with the protocol provided by web browsers in order to download all registered Internet application performance measurements. The tool was created by P. Narasimhan who presented its potential at *Front End Ops* conference in 2014. [11].

The main advantage of the browser-perf tool is the possibility of its integration with Selenium. It is developed on the popular Java platform to perform functional tests of web applications. Owing to the functional tests, the programmer can simulate the action of the user performing actions on the website. The browser-perf application in the calling process requires the selection of the Selenium server instance running together with the WebDriver interface dedicated to each browser separately. In order to conduct tests using the Google Chrome browser, the programmer should indicate the appropriate interface, which is equipped with methods for exchanging data with Google Chrome Debugger Protocol.

Perfjankie. This tool is a program that extends the possibilities to perform automated measurements using the browser-perf application. Owing to evoking the command with indication of the number of repetitions for a given scenario, the collected measurements are saved in the Apache CouchDB database. Based on a series of tests, perfjankie provides a preview of collected measurements in the form of characteristics and diagrams available from the web application level [12].

4 Research Stand

For the purposes of the research three applications launched on the Node.js platform were designed and implemented. These applications were created according to SPA pattern in JavaScript language. The aim of the project was to implement the same functionalities in each application separately, using different architectural approaches. Each of the approaches is transformed according to its own implementation (selected libraries) of the HTML markup language into the object document model. As a result, in two cases, the constructed user interfaces are represented by a virtual DOM and libraries supporting the reactive functional programming – FRP.

The first application was implemented in accordance with the popular architectural pattern of reactive web applications and consists of the React.js, Redux and Redux-Observable libraries. The second one is an implementation using the reactive Cycle.js micro-framework. The third one is a classic implementation using native technologies to build internet applications (no reactivity and VDOM): JavaScript and DOM API.

Each application communicates with services on the network (e.g. external services, micro services, etc.) using the REST API. The webpack tool was used to work on the application. The tool is required, among others, to transpilate the standardized JavaScript code version 6 and newer to the form understood by all web browsers. Webpack allows one to compile many different types of resources (source files) to one result resource (bundle), which improves the performance of modern web applications.

4.1 Application Architecture

React.js, Redux and Redux-Observable. The React.js library is the most popular solution in the network in the field of available VDOM implementations, provides flexibility and a component approach in the process of building interfaces. The advantage is the ever-growing popularity and huge community that develops additional tools and components of universal purpose, known as *third-party* software. The architecture of the SPA application based on React.js is highly scalable, and the control over complexity is ensured by the unidirectional data flow. To manage the state of the application and control the flow of data, the popular Redux library was chosen, which is the most popular (according to github.com) implementation of the *Flux* architecture pattern. The combination of React.js and Redux libraries is a configuration that meets all the requirements set for modern web applications built for the implementation of the front-end layer.

To provide patterns to work according to reactive extensions, the RxJS library was added to the above configuration, which is a dependency for the Redux-Observable library. Redux-Observable is middleware for the Redux library that manages the state of the application. As a result, each time the action is called anywhere in the user interface to change the data model, it triggers the launch of the reactive processing functions that are captured and treated as a data stream. Redux-Observable also extends native methods for handling asynchronous AJAX queries with reactive patterns and functional stream processing operations, e.g. queuing and allowing requests to be cancelled, or it composes complex API queries from multiple individual requests.

Versions of libraries applied: React.js 16.2.0, Redux-Observable 0.17.0, Redux 3.7.2, RxJS 5.5.2. In further considerations this application will be referred to as **RRO**.

Cycle.js. This is one of the first frameworks to develop fully reactive web applications. The factor that distinguishes Cycle.js from other frameworks is its scheme of operation.

Cycle.js concentrates all kinds of user interface events along with rendering views, returning data to views, etc. It brings to mind the circulating data flow (*cycle*) from the input, or source (*source*) to exit (*sink*). The flow is invisible to the programmer – it makes use of reactive patterns and the FRP paradigm.

The data flow takes place between packages (*controllers*), e.g. @cycle/dom, @cycle-history, @cycle/html, @cycle/http, @cycle/isolate, @cycle-run, @cycle/most-run, @cycle/run, @cycle/rxjs-run itd. The @cycle/house package responsible for generating the VDOM structure uses the *snabdom* library.

Package versions used: cycle/house 18.3.0, cycle/http 14.8.0, cycle/run 3.3.0, cycle/time 0.10.1, xstream 11.0.0. In further considerations this application will be referred to as **CLE**.

JavaScript and DOM API. The third of the proposed solutions is the native approach to constructing web applications in the JavaScript programming language. The approach is devoid of reactive features and uses the classic DOM API. The application prepared in the ES6 version (Ecma International) for the purpose of comparing measurements as a reference point in relation to the two previous suggestions. In further considerations this application will be referred to as **JSD**.

4.2   Application Functionalities

In each application, a popular piece of reality was implemented, involving the management of personal data. The main criterion for such a choice is its complexity. The concept of complexity is understood as the number of elements (nested data structures) transferred to the view and the number of generated DOM/VDOM nodes based on them. The element is represented here by the being of an individual.

Another criterion is the number of actions performed by the user and their level of advancement. The concept of an advanced action is understood as its likely impact on the current state of the DOM/VDOM structure, and as a result the number of operations and modifications on its structure that must be performed in order for the action to be successful. This criterion is an abstraction of some kind, as it is impossible to estimate how many operations will be performed by the indicated DOM/VDOM engine implementation as a result of the action. It is, however, possible to predict the probable impact of the action on the behaviour of interface elements and its structure. The visualization of the tree structure of the document, and subsequently the analysis of operations on its nodes (creation, deletion, modification) can be of help in this case.

The following application views (interfaces) were used in the research:

- **List of persons.** The list of all persons is presented in the form of a table. The downloaded data is stored in JSON format. The experiment with the use of this interface will allow to evaluate the page loading speed and to construct an object-based document model based on the number of data transferred.
- **Personal file.** Access to the file is possible through the action of expanding the table row, where each of the rows represents an individual's being. The action evoked in this way changes the interface in the place indicated by the user.

- **Searching for personal data.** For all beings that meet the search criteria, a file with detailed data is displayed. As a consequence, during the experiment it will be possible to measure the results for "costly" DOM/VDOM tree transformation operations. List filtering and search operations are performed using data loaded into the memory.
- **Deleting personal data.** In the scenario of examining this operation, it is planned to measure the results of the removal efficiency of the set number of elements. Items (indexes) of elements on the list will be random – for the first variant. In the other variant of the experiment scenario, it is possible to remove all elements. During the experiments with these operations, the removal performance of nodes from the document model object tree (DOM/VDOM) will be measured.

5   Research

Performance research were planned for all the three applications. Experiment scenarios are specified in Table 2.

Table 2. Research plan for each of the designed application interfaces based on the prepared scenarios.

Experiment	Objective	Parameters
Loading the list of personal data with a specific number	Measurements of page loading times and DOM/VDOM construction based on the amount of transferred data	Lists with lengths of 100, 200, 500, 1000 components
Searching the full list and displaying cards of persons who meet the criteria	Measurements of execution times of complex DOM/VDOM tree transformation operations – creating and attaching nodes when the action is called	Search phrases entered in the filter fields for a list of 200 and 1000 elements
Deleting data of selected persons – a list of people indicated in random order	Measurements of the DOM/VDOM node removal times from the random location of the previously loaded structure	A one-dimensional, unordered table containing random item identifiers to remove from the 200 and 1000 elements list
Deleting all personal data from the previously loaded list	Measurements of removal times of all DOM/VDOM nodes	One-dimensional, ordered table with a length of 1000

Before starting the experiments, the number of generated nodes in the DOM/VDOM tree was checked and unified for the applications implemented according to the three accepted architectural patterns, so that their number was the same. The number of nodes for the interface presenting a specific list of people is shown in Table 3.

Table 3. The number of DOM/VDOM tree nodes generated for the view presenting a list of people of a specific length.

The length of the list of persons	100	200	500	1000
The number of DOM/VDOM nodes	1948	3848	9548	19048

The experiments were carried out on a Dell Latitude E5470 computer with the following parameters:

- processor: Intel Core i5-6440HQ (4 cores, from 2.6 GHz to 3.5 GHz, 6 MB cache),
- memory: 16 GB (SO-DIMM DDR4, 2133 MHz),
- hard drive: 256 GB SSD M.2,
- graphics card: Intel HD Graphics 530,
- operating system: Ubuntu LTS 16.04.

The entire process of collecting (reading) and processing data by the browser-perf tool happens after the end of the Google Chrome Debugger Protocol activity. Directly during the scenario, browser-perf does not read or record any results, so experiments could be carried out on a single computer.

For each scenario, the measurements were performed fifteen times. As it was a small sample, the median was determined based on the registered results. It was noted that the Loading metrics group had disproportionally low values in relation to other groups, therefore it was not included in the analysis of results.

5.1   Loading a List of a Specific Length

The results of the experiment consisting in loading a list of a specific length are shown in Fig. 1.

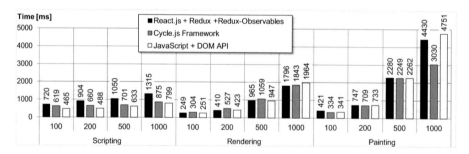

Fig. 1. Diagram of values of group metrics scripting, rendering, painting, for the scenario of reading lists with lengths of 100, 200, 500, 1000 items

The longest waiting time for a full list of items was recorded in the RRO application. The Scripting metric is responsible for this state of affairs, which is the time required to interpret and run JavaScript code. This is due to the complexity of the architecture used to program this version of the application.

The shortest total loading times were recorded for the JSD application for a list consisting of 100 and 200 items. When loading a list of 500 and 1000 elements, the shortest times were recorded in the case of CLE application. This is due to the much shorter times in which the application processes the site layout generation events and develops the CSSOM and DOM rendering tree – the Rendering metric. Shorter waiting times also apply to events responsible for rasterizing and displaying the website via a web browser – the Painting metrics.

It is worth noting that in the case of JSD applications, as the number of DOM nodes increases, the values of the metrics from the Rendering and Painting groups increase significantly. The results show that constructing the virtual structure of the DOM tree is expensive – as indicated by the Scripting metrics, while the time of preparation and display of elements (Rendering and Painting) decreases as their number increases. In the case of applications using VDOM, better results are achieved by CLE using the *snabdom* library.

5.2   Searching the Full List

The results of the experiment consisting in searching a list of a specific length are shown in Fig. 2.

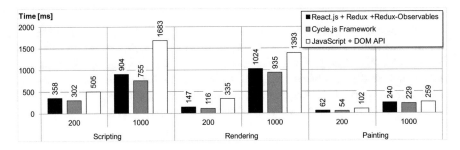

Fig. 2. Scripting, rendering, painting metrics value chart for the search scenario of lists with 200 and 1000 elements

The shortest times for both parameters (200, 1000 elements) were recorded for the CLE application. Similar results were obtained by the RRO application (8–9% longer time of performing operations). Definitely the longest times were observed for JSD applications (for each metrics). The biggest differences can be noticed between the Scripting and Rendering metrics, which reflects the process of constructing and modifying the object document model nodes. Contrary to the previous experiment, reactive applications that use VDOM deal with direct operations of modifying particular nodes much better.

5.3   Deleting Indicated Elements from the List in a Random Order and Deleting All Elements

As in the previous experiment, lists of 200 and 1000 elements were used. In the first variant of the scenario, the indicated elements were removed from the list in a random order. The number of items removed was 100 out of 200 and 500 out of 1000 (in both cases – half of the elements). In the other variant, all elements from the list of 1000 elements were removed. The results of the experiments are shown in Fig. 3.

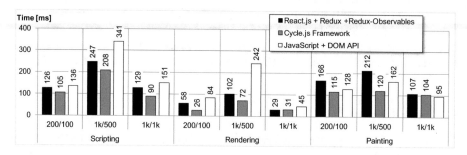

Fig. 3. Scripting, rendering, painting metrics value chart for scenarios for deleting items from the list

In the case of removing items arranged in random order, the CLE application works best. All the metrics indicate this. As in the previous experiment, the reactive applications that use VDOM deal with direct operations of object-oriented document model much better. In the case of removing all items from the list, much lower values of the metrics were observed. Above all, the workload associated with page layout processing and the reactive filtering of the list as application program structure is lower.

6   Summary

The performance studies showed that the architecture of Cycle.js is highly flexible, and the application using it was characterized by the shortest times of processing large amounts of data in a web browser. The case of the website loading scenario with the loading of a multi-element data structure proves that the complexity of the application has a decisive impact on the total time of performing this action. Owing to the low degree of complexity and the volume of the source code Cycle.js allows applications to achieve similar processing times, to applications deprived of the reactive features.

The React.js, Redux, Redux-Observable architecture brings with it considerable burden of the source code, and constructing an alternative structure for storing DOM tree elements in the process of completely loading the site is not an optimal solution. This architecture will not meet non-functional requirements for not very complex applications with low complexity, whose initial response time is important from the end-user perspective. The analysis of the search results, adding and removing HTML nodes showed that applications that use the virtual representation of the DOM tree perform tasks much faster than those that use the DOM in the traditional way.

Issues related to the performance of web applications should be an important factor when making decisions by the system architect today. S/he must carefully choose the right architecture or tools (libraries) at the software specification and design phase. The conducted research may be helpful before taking the key decision. It should be remembered, however, that the architecture beyond performance aspect is also characterized by other features, such as the speed of the implementation process, the ease of testing individual application components, community support, the availability of qualified programmers, etc. Therefore, the decision to choose the right architecture should not be taken just because performance tests have been positive. Before deciding on the choice of architecture, it is necessary to conduct the case study and the analysis of functional requirements. Each choice from the suggested architectural solutions should be dictated by the class of the system and the degree of its complexity. Table 4 lists advantages and disadvantages for each of the architectural solutions suggested with their application (usability) determined. In the authors' opinion, it is the most important result of the research. The review of libraries, documentation support, application design, application implementations, preparation of research scenarios and carrying out the measurements provided a lot of knowledge, thanks to which it was possible to prepare this table.

Table 4. Comparison of advantages and disadvantages of the application of the architectural patterns researched.

	Advantages and disadvantages	Applications
React.js + Redux + Redux-Observable	– meets all the requirements for SPA applications – consists of the most popular libraries in the field of VDOM implementation, and application state management – employs the most popular solution in the field of reactive programming in JavaScript (reactive RxJS extensions) – high scalability, there are many proven patterns of the source code organization – high complexity, relatively long loading times of the first view – efficient management of data flow, short waiting times for data processing operations – large file size after using the lossless gzip data compression method via a web browser	– highly scalable applications (systems) that work with multiple access points – REST API communication with NoSQL databases – mobile applications with a large number of visits, e.g. social networks, navigation using geolocation, multimedia (games and entertainment)

(continued)

Table 4. (*continued*)

	Advantages and disadvantages	Applications
Cycle.js	– fully reactive framework that uses VDOM to create applications that meet the SPA requirements – highly flexible, the ability to choose the appropriate dependencies when working with data streams – low degree of complexity and the volume of the source code – relatively fast times of loading the first view and processing data – weak community support, the project is still in development – low scalability, a problem with the organization of the source code of the application – no support for the extension software, i.e. *third-party* software	– real time low latency and short response times web applications – mobile applications of low complexity, "light" and easy to use – interactive forms, instant messengers
JavaScript + DOM API	– very low scalability, freedom in the organization of the source code, no design framework unsupported by any framework – speed and immediacy of performing operations, – problem with maintaining the source code, difficulty in testing, susceptibility to errors	– web application prototypes, mainly interfaces and their presentation layers

7   Related Works

SPA performance research has been conducted for several years in the Department of Computer Science at the Wrocław University of Science and Technology. In 2015, Karabin and Nowak analyzed the performance of SPA constructed with the help of two JavaScript frameworks: AngularJS and Ember.js. They showed that the size of the framework or library is of no major significance for SPA type web applications, as the operations on the DOM tree and the speed of execution of JavaScript code are the actual bottlenecks [8].

In 2016, Stępniak and Nowak analyzed the effectiveness of methods accelerating the process of loading applications of SPA type, including the mechanisms offered by the HTTP/2 protocol. The results indicate that the technique accelerating the process of loading an SPA application the most is the minification of JavaScript code, which significantly reduces the size of transferred resources. The removal of unnecessary CSS rules, despite the small difference in reducing CSS files, undoubtedly has a positive impact on accelerating an SPA application [14].

In 2017, Markiewicz and Nowak used the existing commercial system based on SPA for research. As a result of the research, it turned out that for the web system and for all systems of similar type, the primarily negative factor in terms of performance is data processing by a browser. The time of server processing is of lesser importance than client processing. The overhead of TCP protocol and DNS mechanism is very small and non-significant [10].

The idea of using a virtual DOM to improve the performance of SPA applications was described in 2013 on a specialized blog by C. Chedeau. The author describes the differential algorithm used in React.js. there As finding the minimum number of modifications between any two trees is a problem O (n^3), React.js only tries to group trees according to level. It drastically reduces the complexity and is not a big loss, as it is very rare to move a component to another level in the tree in web applications. They usually move only laterally among children [4].

References

1. Apparao, V., et al.: Document object model (DOM) level 1 specification. World Wide Web Consortium. MIT Press, Cambridge (1998). https://www.w3.org/TR/1998/REC-DOM-Level-1-19981001/
2. Auchenberg, K.: Our web development workflow is completely broken. Full Frontal 2013. 5th JavaScript Conference. Brighton, UK (2013). https://www.youtube.com/watch?v=ctwEcZC_mmI
3. Basques, K.: Get started with analyzing runtime performance. Tools for Web Developers. Google, Mountain View, USA (2018). https://developers.google.com/web/tools/chrome-devtools/evaluate-performance/
4. Chedeau, C.: React's diff algorithm. Performance Calendar (2013). https://calendar.perfplanet.com/2013/diff/
5. Czaplicki, E., Chong, S.: Asynchronous functional reactive programming for GUIs. In: Proceedings of the PLDI 2013, 34th ACM SIGPLAN Conference, Seattle, Washington, USA, pp. 411–422 (2013)
6. Freed, T.: What is virtual DOM. Specialized blog (2016). https://tonyfreed.blog/what-is-virtual-dom-c0ec6d6a925c
7. Husain, J.: ECMAScript observable. Stage 1 draft. Proposal introduces an observable type to the ECMAScript standard library. Ecma International, Geneva (2017). https://tc39.github.io/proposal-observable/
8. Karabin, D., Nowak, Z.: AngularJS vs. Ember.js—performance analysis frameworks for SPA Web applications [in Polish]. In: Kosiuczenko et al. (ed.) From Processes to Software: Research and Practice/Sci, pp. 137–152. Polish Information Processing Society, Warsaw (2015)
9. Mann, J., Wang, Z.: Performance timeline. W3C recommendation (2013). World Wide Web Consortium. MIT, Cambridge (2013). http://www.w3.org/TR/2013/REC-performance-timeline-20131212/

10. Markiewicz, R., Nowak, Z.: User-perceived performance analysis of single-page web application using navigation, resource and user timing API. In: Kosiuczenko, et al. (eds.) Software Engineering Research for the Practice/Sci. Polish Information Processing Society, Warsaw, pp. 105–121 (2017)
11. Narasimhan, P.: Adding rendering metrics to browser performance. In: Front End Ops Conference (2014). https://www.youtube.com/watch?v=Rl6ZAd_Rd20
12. Narasimhan, P.: Making frontend performance testing a part of continuous integration— PerfJankie. In: Velocity Conference, Santa Clara, USA (2014). http://nparashuram.com/perfslides/
13. Occhino, T., Chen, J., Hunt, P.: Rethinking web app development at Facebook. In: F8 2014 Developer Conference, San Francisco, USA (2014). https://code.facebook.com/videos/242117039324244/rethinking-web-app-development-at-facebook-f8-hacker-way/
14. Stępniak, W., Nowak, Z.: Performance analysis of SPA web systems. In: Borzemski, et al. (eds.) Proceedings of the 37th International Conference on Information Systems Architecture and Technology, ISAT 2016, Pt. 1, pp. 235–247. Springer (2017)
15. Zhiheng, W.: Navigation Timing. W3C Recommendation (2012). World Wide Web Consortium. MIT, Cambridge (1998). https://www.w3.org/TR/navigation-timing/

Real-Time Comparable Phrases Searching Via the Levenshtein Distance with the Use of CUDA Technology

Witold Żorski$^{(\boxtimes)}$ and Bartosz Drogosiewicz

Cybernetics Faculty, Military University of Technology, gen. Witolda
Urbanowicza 2, 00-908 Warsaw, Poland
{wzorski,bartosz.drogosiewicz}@wat.edu.pl

Abstract. The paper presents a real-time method for finding strings similar to a given pattern. The method is based on the Levenshtein metric with the Wagner–Fischer algorithm being adopted. An improvement is proposed to this well-known technique, a histogram-based approach which resulted in significant reduction of calculation time without a noticeable loss of correctness. Additionally, the used Wagner–Fischer algorithm has been massively parallelized with CUDA technology. The presented method is very flexible as one can define a task-suitable vocabulary, even for abstract elements, far beyond applications relevant to alphanumeric objects. The presented approach seems to be promising for networking and security applications as it is suitable for real-time analysis of data streams.

Keywords: Levenshtein metric · Wagner–Fischer algorithm
CUDA technology · Real-time systems

1 Introduction

The Levenshtein distance [13] is a metric in the space of characters (symbols) strings. It has been defined based on the following three simple edit actions: insertion (inserting character into a string), deletion (removing a character from the string), substitution (changing a character in the string). The Levenshtein distance is the smallest number of such simple edits that change one string into another. This can be described as $\mathrm{Lev}(a, b) = \mathrm{lev}_{a,b}(|a|, |b|)$, in the following recursive way:

$$
\mathrm{lev}_{a,b}(i, j) = \begin{cases} \max(i, j) & \text{if } i = 0 \text{ or } j = 0, \\ \min \begin{cases} \mathrm{lev}_{a,b}(i-1,\ j) + 1 \\ \mathrm{lev}_{a,b}(i,\ j-1) + 1 \\ \mathrm{lev}_{a,b}(i-1,\ j-1) + 1_{\text{if } a_i \neq b_j} \end{cases} & \text{otherwise.} \end{cases} \tag{1}
$$

Examples: Lev('kitten', 'sitting') = 3; Lev('abcd', 'dcba') = 4; Lev('golden', 'dongle') = 5; Lev('dictionary', 'indicatory') = 6; Lev('distance', 'instead') = 6.

The Levenshtein distance is a particular case of a more general approach of measuring distance between sets [7, 12], which, in turn, belongs to the field of topology

© Springer Nature Switzerland AG 2019
P. Kosiuczenko and Z. Zieliński (Eds.): KKIO 2018, AISC 830, pp. 135–150, 2019.
https://doi.org/10.1007/978-3-319-99617-2_9

[14]. Some alternatives [10, 16] or modifications [9, 24] to the standard Levenshtein distance are also popular, especially when transposition of two adjacent characters is taken into account (the Damerau–Levenshtein distance[1]).

There are many interesting practical applications of the Levenshtein metric. Recently, a very popular one is the handwriting recognition [4, 23]; another challenging task is prediction of a user behaviour [9]. The method is willingly engaged for databases as a supporting tool for detection of similarities [19] or duplications [8], as a cross-language matching technique [18], or for some errors correction [1]. There are even attempts to use the Levenshtein metric in the area of computer vision, e.g. for gesture recognition [17].

This paper presents a proposal improvement to the standard approach based on the Wagner-Fischer algorithm, particularly if the distance between compared phrases is significant. The described method introduces the use of histogram-like technique what leads to a significant reduction of calculations (by order of magnitude) with a negligible influence on result preciseness. The method was verified within a computer system that allowed to perform pattern matching tasks [6, 15] on transferred (via a network connection) alphanumeric streams of data. It is shown that the method can be massively parallelized (by striping the streams of data) with the use of AMD APP [25] or CUDA [20] technology.

2   Wagner-Fischer Algorithm

A direct implementation of the Levenshtein recursive formula (1) is extremely inefficient [11]. The solution to this problem is the Wagner-Fischer algorithm [22], that enables to perform Levenshtein distance calculation efficiently [2, 3]. The algorithm is based on the observation that the distance for all prefixes of two considered strings can be gradually indexed in a matrix, and the last calculated value will constitute the result. This procedure can be illuminated on an example – Fig. 1 shows the calculated matrix for strings "Sunday" and "Saturday" (no difference between uppercase and lowercase letters is taken into account). For all idx and idy, D(idx, idy) holds the Levenshtein distance between the first idx characters of the first string and the first idy characters of the second string. The figure includes also a useful fragment of a simple (Scilab) code related to the Wagner-Fischer algorithm.

As we are interested only in distances smaller than assumed threshold τ, then it suffices to compute a diagonal stripe of width $2\tau + 1$ in the distances matrix presented in Fig. 1; this way, the algorithm can be run in $O(l_P \tau)$ time, where l_P is the pattern length (see Fig. 2).

---

[1] Damerau worked at IBM on problems of detection and correction of spelling errors [5] before the Levenshtein metric was invented.

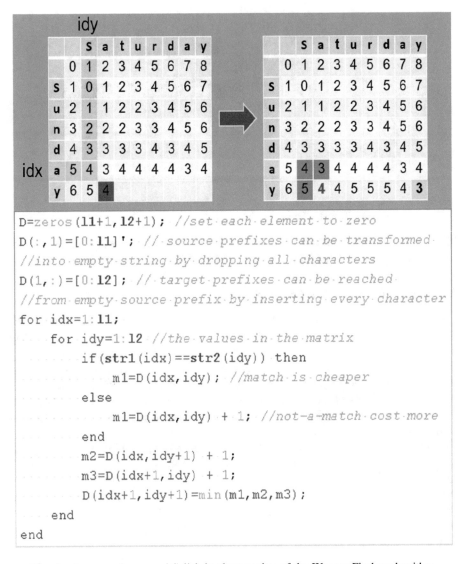

```
D=zeros(11+1,12+1);  //set each element to zero
D(:,1)=[0:11]';  // source prefixes can be transformed
//into empty string by dropping all characters
D(1,:)=[0:12];  // target prefixes can be reached
//from empty source prefix by inserting every character
for idx=1:11;
    for idy=1:12  //the values in the matrix
        if(str1(idx)==str2(idy)) then
            m1=D(idx,idy);  //match is cheaper
        else
            m1=D(idx,idy) + 1;  //not-a-match cost more
        end
        m2=D(idx,idy+1) + 1;
        m3=D(idx+1,idy) + 1;
        D(idx+1,idy+1)=min(m1,m2,m3);
    end
end
```

Fig. 1. An example use and Scilab implementation of the Wagner-Fischer algorithm.

3 The Method

The Wagner-Fischer algorithm is the starting point for the method, and the main assumption is to speed-up the pattern matching task on fragments of transferred streams of data as well as to improve flexibility. The initial task can be formulated in this way (see Fig. 2): to develop a quick search method of "similar" (to a given pattern) fragments in a file (with ASCII, ANSI, or UTF-8 encoding).

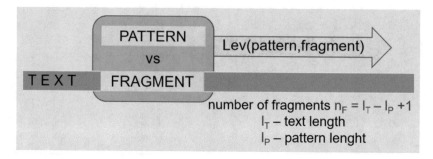

Fig. 2. A conceptual representation of the considered task.

3.1 An Example Simple Task

The example in Fig. 3 reveals the power of the tool (i.e. a Levenshtein based method) as the received result is beyond the human perception.

3.2 An Analysis of a Not So Simple Task

Let us now consider the same pattern "Levenshtein", but the analysed text will be much longer, i.e. the screenplay[2] for "Raiders of the Lost Ark" movie. Such a long alphanumeric string (247901 characters) gives an opportunity to acquire some stochastic data. As presented in Fig. 4, there are two fragments with the smallest distance (equal 5) to the pattern. Much more interesting is a diagram that shows some kind of received distances distribution, what can be find in Fig. 5. We can find out there, that the number of fragments with distance equal 6 is much smaller then with distance equal 7 and so on.

This observation leads to the idea that maybe there is a possibility to perform some kind of low-cost initial verification on fragments, before the computationally complex Levenshtein distance is calculated (using the Wagner-Fischer algorithm) for all fragments. Therefore, if there exists a fast method of preliminary detection of inadequate fragments (comparing to a given pattern), we will be able to significantly decrease the time of calculations (for a transferred stream of data), proportionally to the level of fragments removal.

3.3 The Alphabet and Histograms

Let us consider the following alphabet[3] (26 + 2 + 1 chars): from "a" to "z", {space}, {dot}, and {any other char}. This basic alphabet can be numbered or indexed from 1 to 29. The last element {any other char} is very important as the task of matching patterns is simplified, and extended on some difficult cases.

---

[2] http://www.dailyscript.com/scripts/RaidersoftheLostArk.pdf
[3] We will not distinguish between uppercase and lowercase letters.

Text: *Euler was born on 15 April 1707, in Basel to Paul Euler, a pastor of the Reformed Church, and Marguerite Brucker, a pastor's daughter. He had two younger sisters named Anna Maria and Maria Magdalena. Soon after the birth of Leonhard, the Eulers moved from Basel to the town of Riehen, where Euler spent most of his childhood. Paul Euler was a friend of the Bernoulli family - Johann Bernoulli, who was then regarded as Europe's foremost mathematician, would eventually be the most important influence on young Leonhard. Euler's early formal education started in Basel, where he was sent to live with his maternal grandmother.*

Pattern: *Levenshtein*

Closest fragments: index=454, Lev=7,   *ld eventual*
                      index=456, Lev=7,   *eventually*
                      index=589, Lev=7,   *live with h*

Fig. 3. An example result of searching fragments similar to a given pattern.

Text: {*"Raiders of the Lost Ark"* movie script}, $I_T = 247901$

Pattern: *Levenshtein* , $I_p = 11$, $n_F = I_T - I_p + 1 = 247891$

Closest fragments:

index=205208, Lev=5,   *begins tyin*
            {... *and begins tying himself* ...}
index=216879, Lev=5,   *even stran*
            {... *light is even stranger,* ...}

Fig. 4. Results received within the second considered task.

Now, we can create a graphical representation of the distribution, i.e. histogram, for such numerical data as alphanumeric strings. Figure 6 presents histograms for example strings "A. Einstein" and pattern "Levenshtein".

Let us assume that we have fragment histogram H_F and pattern histogram H_P. These distributions allows us to calculate the histogram distance (based on L_1-*norm*) between considered objects using the following formula [21]:

$$H_d = \frac{1}{2} \sum_{i=1}^{N} |H_F(i) - H_P(i)|. \tag{2}$$

In the example presented in Fig. 6 the histogram distance between the fragment and the pattern is equal 4. and the Levenshtein distance is 5.

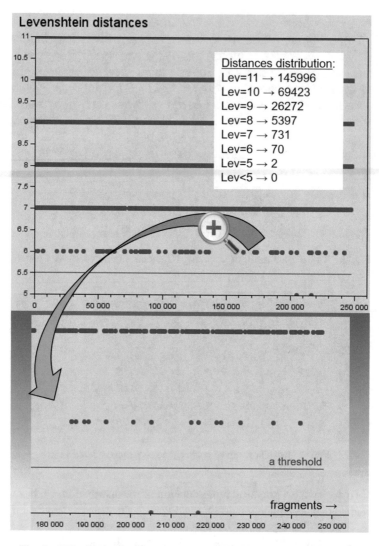

Fig. 5. "Distribution" of fragments distances for the considered pattern.

3.4   Checking Neighbours

It is quite easy to find out (from the Levenshtein metric definition) that the maximum possible distance between neighbour fragments is 2. On this basis a simple improvement to the method can be introduced. If an analysed fragment is distant enough from the pattern the next fragment can be considered as not good enough to be taken into account. The second next fragment can be also considered the same way but now the maximum possible distance is 4; this can be accepted only if there is a distance reserve and the pattern is long enough.

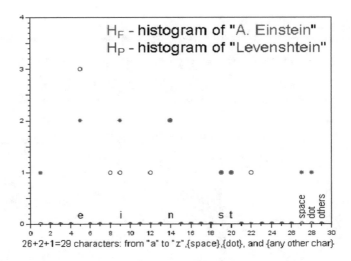

Fig. 6. Histograms for strings: "A. Einstein", "Levenshtein".

This improvement was examined (see Fig. 15), it works only with longer patterns, and unfortunately is sometimes faulty (the threshold problem). Nevertheless, it is possible to use this approach as a supporting technique.

3.5 The Histogram Analysis

In the case of histogram analysis [26], only the histogram for the first fragment has to be obtained in a standard way, and for subsequent fragments histograms are determined by modifying the predecessor (by removing the first character and adding a new one), which is very important and proved to be profitable.

PROCEDURE for histogram analysis:

Step 1: Determine histogram of the pattern.
Step 2: Determine histograms for all fragments (one by one).
Step 3: Compare received fragments' histograms with the pattern histogram using formula (2).
Step 4: Remove all fragments (by marking them) with the histogram distance greater than a histogram threshold; they will be excluded when calculating the Levenshtein distance.

The sensitive element in this analysis is the histogram threshold. Based on several experiments it turned out that it should be greater than half the length of the pattern, in order to not cause loss of correctness.

This simple method significantly reduces time of the calculations for the whole process (often more than 50%), what is visible in Fig. 7.

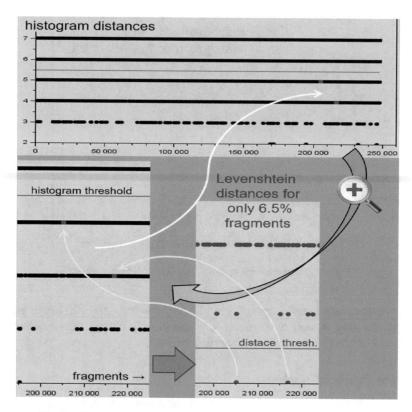

Fig. 7. The relation between histogram and the Levenshtein distances.

4   The Verification Environment

The method was verified in a system presented in Fig. 8; it consists of two x86 computers (Intel Core i5), equipped with 8 GB of RAM, a Fast Ethernet/GbE adapter (100/1000 Mb/s), and a GPU supporting CUDA technology.

Such a testing environment allows to perform high-performance real-time experiments on a stream of data transferred with two standard Ethernet speeds. The stream of alphanumeric characters can be transferred between PC1 and PC2, while a sniffer-like software (running on any computer) performs pattern matching task on the receiving data stream. The task performed in the system can be calibrated for patterns of various length in order to detect limits for the elaborated method.

The elaborated method was first prototyped with Scilab and then translated to C++ to gain at least several times performance improvement, as Scilab is a script language (a command line interpreter) and C++ source code is being compiled.

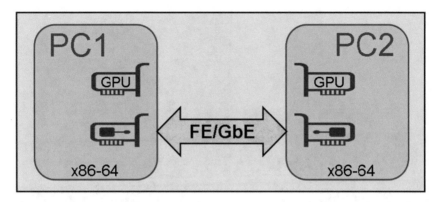

Fig. 8. The verification system: two PCs with suitable network adapters.

5   The Use of CUDA

5.1   The CUDA Technology

The CUDA[4] technology appeared in 2007 as a result of new Nvidia's GPUs branded GeForce 8. CUDA gave program developers direct access to the virtual instruction set and memory of the parallel computational elements in GPUs.

CUDA is a parallel computing platform and programming model [27, 28] that makes using a GPU for general purpose computing simple and elegant. The first widely used architecture was Fermi. Each Fermi streaming multiprocessor (SM) included 32 CUDA cores with ability to perform integer or floating operations simultaneously (see Fig. 9). At present, there are two main CUDA architectures available: Pascal[5] and Volta[6]. The Pascal architecture is very popular and commonly used, while the Volta architecture is extremely expensive, being still available only for professional HPC systems.

5.2   Wagner-Fischer Algorithm Implementation

The Wagner-Fischer algorithm (see Fig. 1) can be performed in a parallel way using the CUDA technology. We will consider two implementations.

The first implementation (see Fig. 10) is direct. It means that the Wagner-Fischer algorithm was directly implemented as a kernel without any optimization.

For the second "optimized" approach it was necessary to construct a new traversed form of array (presented in Fig. 11) in which a single row of cells can be computed in a way suitable for the CUDA execution model.

Assuming that pair (i, j) describes the position of a given element in the Wagner-Fischer array, all elements can be transferred to the traversed array using the

---

[4] Compute Unified Device Architecture

[5] https://docs.nvidia.com/cuda/pdf/Pascal_Tuning_Guide.pdf

[6] https://docs.nvidia.com/cuda/pdf/Volta_Tuning_Guide.pdf

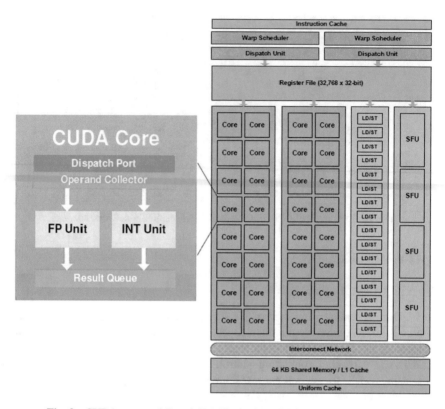

Fig. 9. CUDA core and Fermi SM (Streaming Multiprocessor) structures.

dependency: $(i, j) \rightarrow (i + j, i)$. Figure 11 suggests the way of computing elements of the traversed array on a CUDA device, as every element (i.e. from e_{11} to e_{46}) can be computed based on three previous cell values (greenish in Fig. 11). This requires rows to be computed in sequence, what is presented in Fig. 12. To ensure that rows are computed in desired way, kernels are constructed to contain one block of threads corresponding to rows in the traversed array, as order of their execution could not be guaranteed due to the unknown distribution of blocks among the streaming multiprocessors.

Figure 13 presents the CUDA kernel code. The shared memory is used to store and next swiftly access elements from two previous rows. The rest of the code represents a typical Wagner-Fischer algorithm. If two compared characters match, the diagonal value is simply used as the result; otherwise, the minimum value (from three considered values) is incremented by one. Finally, the result is stored in the global memory to be used by the following kernels. For the last kernel the thread will identify itself and assign result to the specified variable.

```
14    __global__ void WagnerFischerKernel(char * s1, char * s2, int WORDS_NUM, int MAX_LENGTH,
15                                          int * wfmatrix, int * length) {
16        int id = threadIdx.x + blockIdx.x * blockDim.x;
17        int m = 0; int n = 0; int k = 0;
18
19        //Count lengths of strings thread is comparing
20        while (s1[id * MAX_LENGTH + k++] != 0) m++; m++; k = 0;
21        while (s2[id * MAX_LENGTH + k++] != 0) n++; n++;
22
23        //Initialize Wagner-Fischer array
24        int offset = id * MAX_LENGTH * MAX_LENGTH;
25        for (int i = 0; i < m; i++) wfmatrix[offset + i] = i;
26        for (int j = 0; j < n; j++) wfmatrix[offset + j * MAX_LENGTH] = j;
27
28        for (int y = 1; y < n; y++)
29            for (int x = 1; x < m; x++)
30                if (s1[id * MAX_LENGTH + x - 1] == s2[id * MAX_LENGTH + y - 1]) {
31                    int diagonal = wfmatrix[offset + (x - 1) + (y - 1) * m];
32                    wfmatrix[offset + x + y * m] = diagonal;
33                } else {
34                    int diagonalVal = wfmatrix[offset + (x - 1) + (y - 1) * m];
35                    int leftVal = wfmatrix[offset + (x - 1) + y * m];
36                    int upperVal = wfmatrix[offset + x + (y - 1) * m];
37
38                    int min = upperVal < diagonalVal ? (upperVal < leftVal ? upperVal : leftVal) :
39                        (diagonalVal < leftVal ? diagonalVal : leftVal);
40
41                    wfmatrix[offset + x + y * m] = min + 1;
42                }
43
44        //Save the result
45        length[id] = wfmatrix[offset + m * n - 1];
46    }
```

Fig. 10. The first (direct) CUDA kernel implementation (compare Fig. 1).

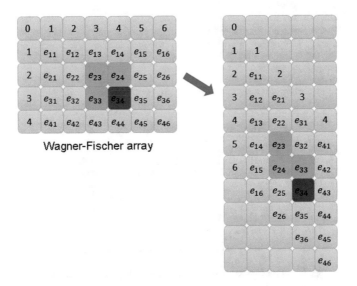

Fig. 11. Conversion of the standard Wagner-Fischer array into a traversed form.

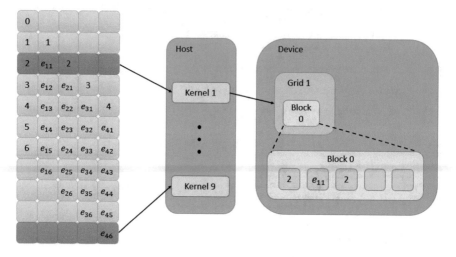

Fig. 12. The use of CUDA execution model.

```
43     __global__ void WFPyramidKernel(int pLevel, char * s1, char * s2,
44           int width, int height, int * wfpyramid, int * lvLength) {
46           __shared__ int previousRow1 [1024];
47           __shared__ int previousRow2 [1024];

49           int i = pLevel * width + threadIdx.x;
50           int x = (i % width); int y = (i / width);

52           previousRow1[x] = wfpyramid[x + (y - 1) * width];
53           previousRow2[x] = wfpyramid[x + (y - 2) * width];
54           __syncthreads();

56           //Only cells equal to -1 require computation
57           if (wfpyramid[i] != -1) return;

59           //Indexes of letters the thread is comapring
60           int s1i = (y - x) - 1; int s2i = x - 1;

62           int result = 0;
63           if (s1[s1i] == s2[s2i]) {
64               int diagonal = previousRow2[x - 1];
65               result = diagonal;
66           } else {
67               int diagonalVal = previousRow2[x - 1];
68               int leftVal = previousRow1[x];
69               int upperVal = previousRow1[x - 1];
70               int min = upperVal < diagonalVal ? (upperVal < leftVal ? upperVal : leftVal) :
71                   (diagonalVal < leftVal ? diagonalVal : leftVal);
72               result = min + 1;
73           }
75           wfpyramid[i] = result;
76           if (i == height * width - 1)
77               *lvLength = result;
78     }
```

Fig. 13. The second (optimized) CUDA kernel implementation.

5.3 Massively Parallel Processing

It proved that the Wagner-Fischer algorithm itself is not very suitable for the CUDA technology. The first algorithm is even faster than the second one for relatively short phrases (with the length below 20).

The main gain of the CUDA technology comes from the possibility of running many Wagner-Fischer algorithm instances in a massively parallel way. In the proposed approach (see Fig. 14) threads are used for series of neighbour fragments and the grid level index (i.e. blockIdx.x) is used to address data stream stripes. Thus, engaging more GPU multiprocessors a striped stream of data can be processed faster and faster.

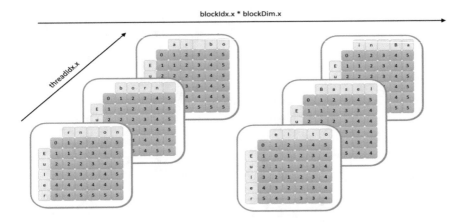

Fig. 14. The use of CUDA parallelism.

6 Results

The first important issue is the gain from the new approach. In order to obtain some objective results a series of experiments was conducted on relatively long (~ 100 KB) alphanumeric strings (random Internet pages and documents). The set of patterns was prepared the way to examine a variety of sentences (up to 25 characters). The received (averaged) results are presented in Fig. 15. The improved method is efficient, **about 5×faster**, especially for longer patterns.

The second significant result comes out directly from experiments conducted in the environment visible in Fig. 8, where relatively very long text streams (~ 100 MB) were transferred. The received data (no CUDA acceleration) are collected in Table 1. The acceleration ratio made by the use of GPGPU strongly depends on the available number of streaming multiprocessors, as the data stream can be buffered and divided into stripes to be massively parallel processed.

Relative calculation time

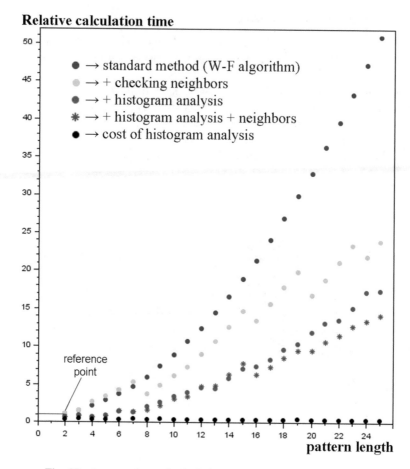

Fig. 15. A comparison of calculation times for considered methods.

Table 1. Bitrates [KB/s] for examined variants of the method (compare Fig. 15).

Method	Pattern length			
	5	10	15	20
Standard	2104	554	275	125
Checking neighbours	2004	1046	438	312
Histogram analysis	7296	1512	956	486
Histogram analysis + neighbours	6985	1686	1067	608

7 Conclusion

The contribution of the presented method to the theoretical work comes from the histogram analysis and is visible when the analysed data stream comprises phrases that are mostly a large edit distance from the pattern phrase. If this distribution pattern exists, it can be successfully exploited.

The proposed method was tested not only in the case of a network connection between computers; thanks to the CUDA technology being adopted, it was successfully examined as a sniffer tool for common data sources, such as mass storage and some standard input devices. Obtained results seem to be quite satisfying and the method is almost fast enough for the Fast Ethernet (100 Mbps). Engaging enough powerful GPU (e.g. GTX 680) it is possible to receive a level of GbE speed. Nevertheless, authors are aware of some issues that can be improved or examined. The very important one is the histogram threshold (this is also the matter of method correctness), as it can be automatically estimated (in a specific implementation) on random samples and should be corrected for long (>20 characters) patterns. The second idea is to use a vocabulary based compression in the case of frequently encountered strings. At the same stage of pre-processing, the transferred stream should be "standardized" to avoid duplications as: (at,@), (dot,.).

As was mentioned, the method can be generalized on any "stream" of objects, as there is a possibility to create a suitable alphabet. Particularly, there is no problem to adopt the method to any Indo-European language, and authors successfully adopted and examined the method for Polish language.

References

1. Abdel-Ghaffar, K.A.S., Paluncic, F., Ferreira, H.C., Clarke, W.A.: On Helberg's generalization of the Levenshtein code for multiple deletion/insertion error correction. IEEE Trans. Inf. Theory **58**(3), 1804–1808 (2012)
2. Andoni A., Onak K.: Approximating edit distance in near-linear time. In: Proceedings of the Forty-First Annual ACM Symposium on Theory of Computing, pp. 199–204. ACM (2009)
3. Backurs A., Indyk P.: Edit distance cannot be computed in strongly subquadratic time (unless SETH is false). In: Proceedings of the Forty-Seventh Annual ACM Symposium on Theory of Computing, pp. 51–58. ACM (2015)
4. Chowdhury, S.D., Bhattacharya, U., Parui, S.K.: Online handwriting recognition using Levenshtein distance metric. In: 12th International Conference on Document Analysis and Recognition (ICDAR) (2013)
5. Damerau, F.J.: A technique for computer detection and correction of spelling errors. Commun. ACM **7**(3), 171–176 (1964)
6. Dong, J., Liu, H.: Semi-real-time algorithm for fast pattern matching. IET Image Proc. **10**(12), 979–985 (2016)
7. Fujita, O.: Metrics based on average distance between sets. Jpn. J. Ind. Appl. Math. **30**(1), 1–19 (2013)
8. Gaikwad, S., Bogiri, N.: Levenshtein distance algorithm for efficient and effective XML duplicate detection. In: International Conference on Computer, Communication and Control (IC4), pp. 1–5 (2015)
9. Harish Kumar, B.T., Vibha, L., Venugopal, K.R.: Web page access prediction using hierarchical clustering based on modified Levenshtein distance and higher order Markov model. In: IEEE Region 10 Symposium (TENSYMP), pp. 1–6 (2016)
10. Kim S.-H., Cho H.-G.: Position-restricted approximate string matching with metric Hamming distance. In: IEEE International Conference on Big Data and Smart Computing (BigComp), pp. 108–114 (2017)

11. Konstantinidis S.: Computing the Levenshtein distance of a regular language. In: IEEE Information Theory Workshop (2005)
12. Levandowsky, M., Winter, D.: Distance between sets. Nature **234**(5323), 34–35 (1971)
13. Levenshtein, V.I.: Binary codes capable of correcting deletions, insertions, and reversals. Sov. Phys. Dokl. **10**, 707–710 (1966)
14. Nagata J.: Modern General Topology. North-Holland Mathematical Library, 3rd edn. North-Holland, Amsterdam (1985)
15. Navarro, G.: A guided tour to approximate string matching. ACM Comput. Surv. (CSUR) **33**(1), 31–88 (2001)
16. Nemmour, H., Chibani, Y.: New Jaccard-distance based support vector machine kernel for handwritten digit recognition. In: 3rd International Conference on Information and Communication Technologies: From Theory to Applications, pp. 1–4 (2008)
17. Nyirarugira, C., Choi, H.-R., Kim, J.Y., Hayes M., Kim, T.Y.: Modified Levenshtein distance for real-time gesture recognition. In: 6th International Congress on Image and Signal Processing (CISP), pp. 974–979 (2013)
18. Medhat, D., Hassan, A., Salama C.: A hybrid cross-language name matching technique using novel modified Levenshtein distance. In: Tenth International Conference on Computer Engineering and Systems (ICCES), pp. 204–209 (2015)
19. Shao, M.-M., Qian, D.-M.: The Application of Levenshtein algorithm in the examination of the question bank similarity. In: International Conference on Robots and Intelligent System (ICRIS), pp. 422–424 (2016)
20. Skłodowski, P., Żorski W.: Movement tracking in terrain conditions accelerated with CUDA. In: Proceedings of the Federated Conference on Computer Science and Information Systems, pp. 709–717 (2014)
21. Cha, S.-H., Srihari, S.N.: On measuring the distance between histograms. Pattern Recogn. **35**(6), 1355–1370 (2002)
22. Wagner, R.A., Fischer, M.J.: The string-to-string correction problem. J. Assoc. Comput. Mach. **21**, 168–173 (1974)
23. Putra, M.E.W., Supriana, I.: Structural offline handwriting character recognition using Levenshtein distance. In: International Conference on Electrical Engineering and Informatics (ICEEI) (2015)
24. Yujian, L., Bo, L.: A normalized Levenshtein distance metric. IEEE Trans. Pattern Anal. Mach. Intell. **29**(6), 1091–1095 (2007)
25. Zhu, H., Cao, Y., Zhou, Z., Gong, M.: Parallel multi-temporal remote sensing image change detection on GPU. In: IEEE 26th International Parallel and Distributed Processing Symposium Workshops and PhD Forum (IPDPSW) (2012)
26. Żorski, W.: The hough transform application including its hardware implementation. In: Advanced Concepts for Intelligent Vision Systems: Proceedings of the 7th International Conference, Lecture Notes in Computer Science, vol. 3708, pp. 460–467. Springer (2005)
27. NVIDIA, CUDA C Programming Guide, March 2018, PG-02829-001_v9.1. https://docs.nvidia.com/cuda/pdf/CUDA_C_Programming_Guide.pdf
28. NVIDIA, CUDA C Best Practices Guide, March 2018, DG-05603-001_v9.1. https://docs.nvidia.com/cuda/pdf/CUDA_C_Best_Practices_Guide.pdf

Smart City Traffic Monitoring System Based on 5G Cellular Network, RFID and Machine Learning

Bartosz Pawłowicz, Mateusz Salach$^{(\boxtimes)}$, and Bartosz Trybus

Rzeszow University of Technology, Rzeszow, Poland
{barpaw, m.salach, btrybus}@prz.edu.pl

Abstract. Smart city is well-known concept of nowadays urban management. With advantages of 5G network, RFID transponders and cloud infrastructure it is possible to create a city traffic monitoring system which helps drivers to reach destination in an optimal time. The system will redirect the driver to specific track to avoid traffic jams and to minimize fuel consumption. In case of an electric vehicle the system will take into account its range and battery power level. The paper presents an idea of such traffic management system. Its components, including 5G communication, RFID-based parking space monitoring and cloud services for supervisory control and machine learning are also characterized.

Keywords: Smart city · RFID · 5G network · Cloud computing
Machine learning

1 Introduction

Currently the term "smart city" is widely used when one wants to describe modern management of a municipal area. Most often the term refers to activities and solutions that involve present day IT technologies to improve life quality of citizens and to introduce flexible and quick adaptation of community responsibilities to their needs.

One of the main tasks that municipal authorities are responsible for is traffic management. From a citizen's point of view this aspect belongs to the most recognizable and crucial problems keeping in view day-to-day experiences. Since traffic management requires to take into account several aspects, e.g. public transport of different types, street network or car parks, it is usually quite hard to establish a flexible solution that will cover all the scenarios that happen on the streets.

Here we present a concept of modern traffic monitoring system for a city. The solution uses recent techniques to make it flexible and adjustable to local conditions. Particularly, the system is able to reorganize the traffic to react to different situations on the streets. The system utilizes 5th generation wireless systems (5G) as the main communication channel and Radio-frequency identification (RFID) to track occupancy of parking space. Microsoft Azure cloud services are employed to implement supervisory control of the traffic and to store and analyse data that describes the current traffic state.

P. Kosiuczenko and Z. Zieliński (Eds.): KKIO 2018, AISC 830, pp. 151–165, 2019.
https://doi.org/10.1007/978-3-319-99617-2_10

The proposed system tracks vehicles that move across the streets, road accidents, road construction, traffic jams, etc. The system is able to guide the cars in an optimal way according to current situation, either by sending guidelines to the driver via a satnav, dynamic traffic signs or a mobile app, or automatic redirection of an autonomous car to a more convenient route.

The cloud is able to determine current traffic state and road conditions using current data that is sent by vehicles online. Some characteristics of the vehicles are taken into consideration, such as its type (lorry, passenger car, public bus), weight and drive (petrol, electric) or drive range (especially important in case of electric cars). Local traffic rules are also considered, such as limited traffic areas, e.g. access depending on daytime, drive type, etc. Drivers are able to point their destination using a mobile app or a dashboard built-in facility and then the system will propose the best route. This way the traffic can be evenly distributed and the driver will reach the destination in an optimal way, possibly avoiding traffic jams. RFID - and cloud-based car park management will guide the driver to a free car park, thus eliminating unnecessary searching for a parking space.

Hence, the main goal of the paper is to formulate the concepts, to design and implement an intelligent road traffic management system using solutions based on RFID technology and cloud computing. The system is assumed to be low-cost, and it may be implemented medium-size and large cities.

2   Related Work

The field of smart cities and intelligent traffic management is currently an important research area. The solutions are often classified by the authors in the field of Internet of Things (IoT). The paper [1] gives a general overview of such approach. Communications is one the main challenges. The authors of [2] consider massive data traffic for IoT applications over a 5G network. Multiple accessing schemas are presented for managing the overall network traffic. An efficient estimation technique for managing the traffic is proposed. To build a smart city it takes wireless networks, telecommunication networks, sensors (GPS, RFID, laser), smart devices which stay connected to each other and communicate to take necessary action. The paper [3] shows the challenges related to multiple devices communicating with each other. It also presents a case study in monitoring green areas being a smart garden system managing conditions and needs of green areas in real-time. The authors of [4] present an IoT-based system for smart cities, called SIGINURB. It allows to develop a range of new services to students, employees, businesses and public administration in the University of São Paulo.

The current status of smart systems for cities in Japan is presented in [5]. It also presents concepts in privacy management platforms, i.e. centric, distributed and hybrid models. The Smart Privacy Platform is positioned at the center of a smart city, associated with other smart systems, and can make use of the new intelligence of cities resulting from the increasingly effective combination of digital telecommunication networks, ubiquitously embedded intelligence, sensors and tags, and software.

Traffic management system is considered as one of the major dimensions of a smart city. The paper [6] presents an application which utilizing vehicular networks to detect traffic condition of the roads. Then electronic and mechanical techniques are used to increase or decrease the number of lines for congested and non-congested sides of a highway. A smart traffic management system is proposed in paper [7] to tackle various issues for managing traffic on. The system takes traffic density as input from cameras and sensors. Artificial intelligence is used to predict the traffic density to minimize the traffic congestion. RFID-based solution is used to prioritize emergency vehicles (e.g. ambulances). The paper [8] also involves RFID, but it focuses on cameras capturing the traffic condition, particularly on detection of traffic violation. The authors of [9] show how the technologies such as RFID or UWB (Ultra-wide band) can be used for providing city-wide navigation and control. Using RFID for road traffic monitoring is also the subject of [10]. An intelligent platform for vehicular control PCIV for traffic monitoring is presented. RFID and cloud computing are applied in public transportation systems. The experimental validation was conducted in a university campus and a small city. On the other hand, the authors of [11] are using iBeacon technology in the field of public transportation to deliver the appropriate advertisements to passengers of public transportation or anyone near the stations.

The paper [12] presents a low-cost parking space management system which uses wireless sensors (magnetic, light, temperature, ambient temperature) to detect presence of a vehicle. The detection is fairly good, but it fails to detect carbon-fiber cars. The parking space information is sent to the main server via MQTT transport, similarly as in the solution presented in this paper. In another paper [13], RFID and OCR are used for automated parking management [14] presents secure and smart parking monitoring, controlling and management solution based on the integration of Wireless Sensor Network (WSN), Radio Frequency Identification (RFID), Adhoc Network, and Internet of Things (IoT). The authors focus on cyber security issues and adopt a lightweight cryptographic algorithm that meets IoT device requirements in term of computational cost and energy consumption. Fog computing has been adopted to process sensitive data. RFID infrastructure is also used in [15]. Location Aware Public Personal Diversity of Information Services based on embedded RFID platform to integrate existing RFID systems and digital information content services is proposed.

3   Architecture of the System

The generic architecture of the traffic monitoring system is presented in Fig. 1. It consists of in-car intelligent modules which communicate with the cloud infrastructure via 5G cellular network.

Using a common cellular network for communication with the cloud and for data transfer seems a big advantage. Apart from the system-specific purposes, the vehicles obtain wireless access to other Internet services and information channels. From this perspective, the only part of the hardware set-up of the vehicle required for the operation of the system are onboard modules that communicate with cloud-based software services. Also, the road must be equipped with measurement point

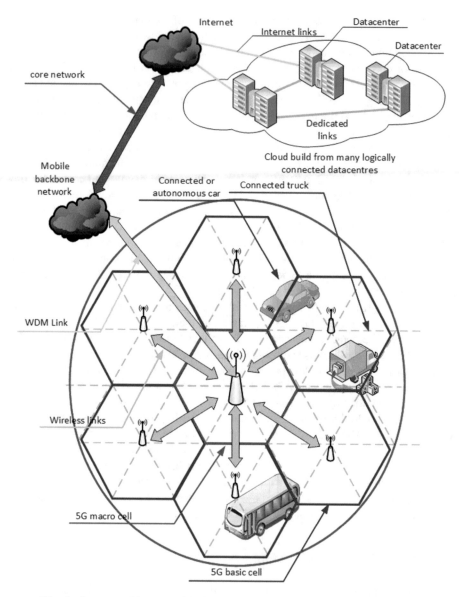

Fig. 1. System architecture with the cloud, connected vehicles and 5G network

infrastructure so that the vehicles may report traffic conditions according to the distance driven.

The core of the system is located in the cloud. It gathers data transferred from the vehicles, processes it and shares the results back to the vehicles. The system uses cloud services to maintain current information about city traffic reported by vehicles. Computational capabilities of the cloud are suitable to store the incoming data in a database, to analyse the traffic and to report unusual situations, such as accidents, road construction, jams, etc. The system is able to instruct the drivers and vehicles to perform some actions depending on traffic conditions.

As shown in Fig. 1, the system is aware of types of vehicles it handles. Therefore, the instructions from the system are adequate to the type. For example, a lorry will be directed to the street which feet its weight, while a city bus or an electric car can use the streets which are closed for other type of vehicles.

4 Traffic Analysis

As stated in Sect. 1, vehicles send traffic data to the system services located in the cloud. By using the data, it is possible to track and visualise their movements and get information about streets or roads they move on. Assuming a car travels from the point A (start) to the point B (destination), one can calculate an optimal route. In the presented system (Fig. 1), we employ machine learning to take into consideration additional factors, such as busy hours, weather or accidents on the route.

The system uses 5G cellular network for communicating with cloud services. The measurement points located in the road infrastructure use RFID technology. Thanks to RFID, it is not required to use GPS for localization. This is particularly important in the areas where GPS is not accessible, such as tunnels, multi-level parking lots, multi-level intersection of streets, etc.

The system uses the collected data to predict the optimal path and to redirect the vehicle to take a roundabout route if a situation requires so (Fig. 2).

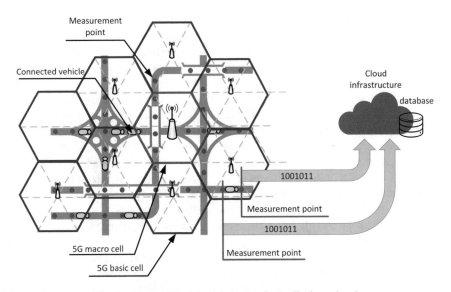

Fig. 2. Route roundabout in case of a traffic jam ahead

The process of navigation proceeds as follows. A driver sets its destination point and requests for navigation route. The cloud receives the request from the vehicle with data important for routing algorithm. To establish a proper route for a car, extra factors are considered, including fuel consumption, time to reach the destination, daytime, weight of the vehicle or the slope of the road. The calculation also involves the vehicle type, traffic intensity or access to side roads. Permissions to enter particular streets or places are also taken into account.

The routing algorithm is trained by using machine learning form of AI to give the best results in terms of the optimal path [16]. The algorithm is constructed in a way that a car will not be redirected to a distant roundabout if the travelling time is longer than the time spent in the traffic jam (Figs. 3 and 4). At the beginning the learning model performs predictions based on raw data set in the database. Every time the algorithm calculates a route, it uses and upgrades its road patterns to set the best route for specified vehicle. This allows to predict better routes to redirect the vehicle and set new path faster each time it calculates a route.

As seen in Fig. 4, the algorithm respects routes for an electric car, such as streets it is allowed to enter or a city district it can access. As the result, the route for an electric car may by different then in case of a vehicle with petrol engine.

The presented system is able to use traffic lights, traffic signs or road information panels to control car flow and to evenly distribute the traffic. By analysing data sent by the vehicles, road accidents are detected and proper action is taken, such as call emergency services or clear a way for them (Fig. 5).

Real-time communication with data sharing between vehicles and the cloud is possible via the 5G network (Fig. 1). We have considered three solutions of two-way communication between a car and the cloud. The first one involves an autonomous car with automatic drive and navigation system. In such scenario, the computer of the car sends the data continually to the cloud through 5G infrastructure. The data contains information about the time of reaching subsequent points of the route. Figure 6 shows data frames sent by a vehicle upon a start and after a predefined time period. Using this data, the cloud service tracks the car and may send back an updated route, which is taken upon request. By collecting information from many vehicles, it is possible to establish a proper driving route to avoid traffic or obstacles on the road. With live data transmission between vehicles and cloud on each measurement point the algorithm can redirect a group of cars on different track to maintain flow on a road.

In case of a non-autonomous car, a navigation application running on a mobile platform (such as smartphone or a tablet) is involved. The application will be provided with the optimal route to the destination according to the current traffic. It is worth noticing, that the route will be send on-line, so it may be updated when the traffic situation changes.

Another solution involves a specialized device mounted permanently to the car. A few hardware boards have been considered for this purpose. In the prototype solution we have used Raspberry Pi 3 model B+ platform with HMI display to run a dedicated application which connects to the cloud, sends and receives data and informs the driver about the route the car should take. Model B+ has been chosen for the final project due to its internal dual band WiFi (2,4 and 5 GHz), BLE 4.2 and Power over Ethernet capability which is needed in some scenarios.

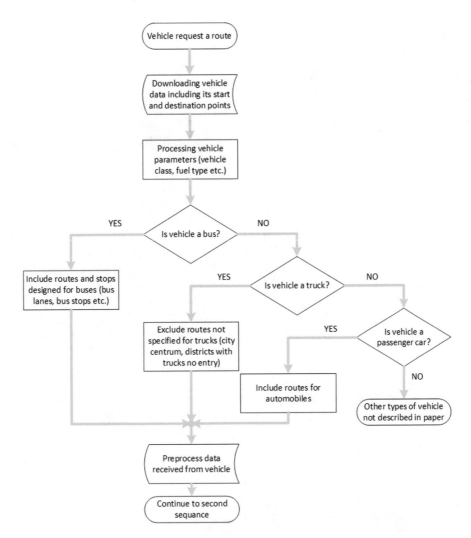

Fig. 3. First phase of vehicle routing algorithm from point A (start) to point B (destination): vehicle class detection.

5   RFID in Traffic and Car Park Management

Communication in the era of intensive development of the road and motorway network, with the ever-increasing volume of transport and road traffic, there is a justified need to control and manage this traffic, and in the longer term - the need to register vehicles having access to protected areas. The last-mentioned needs arise from the fact of intensive formation of separate areas of controlled access, i.e. some of the cities made available only for public transport vehicles or special zones available only for a selected group of rescue or technical vehicles.

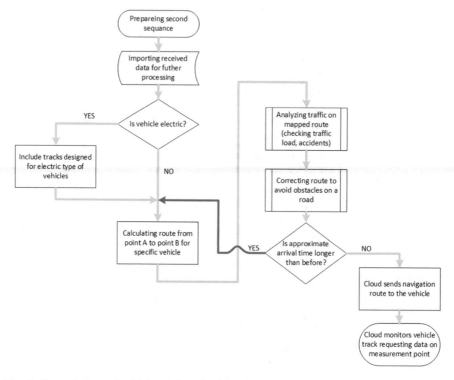

Fig. 4. Second phase of vehicle routing algorithm from point A (start) to point B (destination): creating a navigation route for vehicle with special treatment of electric vehicles

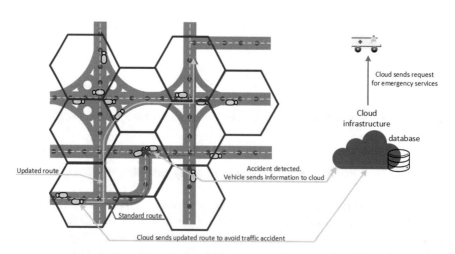

Fig. 5. Calling emergency services in case of a traffic accident

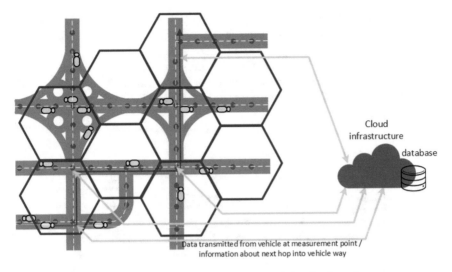

Data transmitted from vehicle at measurement point /
information about next hop into vehicle way

Fig. 6. Communication between a car and the cloud services

The goal of any undertaking aimed at implementing traffic management processes is the selection and implementation of a system that will ensure objective and reliable, static and dynamic identification of vehicles, enabling their simultaneous tracking in real time. Formulated assumptions indicate the necessity to use technology of contactless identification of objects, which will be used in the area of vehicle marking and conducting the process of their automatic identification, both by handheld readers, as well as stationary RFID systems. The implemented activities should be consistent with the essence and legal regulations that apply to the processes of automatic identification of vehicles in the area of transport and road traffic.

Stable operation of the RFID system operating in the traffic management processes area is conditioned by the maintain of the parameters of the components of these systems [17, 18]. The correct functioning of the system is significantly influenced by the type of RFID readers used by management system, their antenna circuits, as well as the type, location and type of used transponder and prevailing environmental conditions. Precise specification of the range of changes in electrical, field and communication parameters characterizing the RFID reader - transponder communication system, enables accurate determination of the area of correct operation of the RFID system. In the proposed system authors decided to implement HF RFID system. The choice was made on the basis of easiness of configuration and price of the system.

In the case of devices working in HF bandwidth, the most popular protocols were standardized in documents ISO/IEC 14443 and ISO/IEC 15693. The first one covers proximity coupling system used in proposed solution, characterized by a greater data bitrate and transmission safety. It consists of four documents:

- ISO/IEC 14443-1: Physical characteristics - defines physical characteristics of Proximity Integrated Circuits Card (PICC) [20];

- ISO/IEC 14443-2: Radio frequency power and signal interface - describes characteristics of a radio interface used for energy transmission and communication between Proximity Coupling Devices (PCD) and PICC [21];
- ISO/IEC 14443-3: Initialization and anti-collision - specifies transponder initialization, data format, time dependences during initialization and anti-collision [22];
- ISO/IEC 14443-4: Transmission protocol - describes half duplex communication protocol [23].

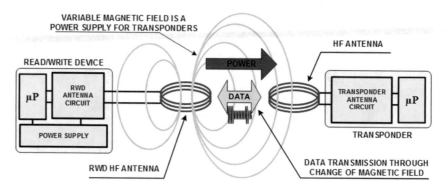

Fig. 7. Energy transmission in HF RFID system

Transponders compatible with ISO/IEC 14443 standard exist in A and B variants, that differ by a method of modulation and coding. The unique feature of all RFID systems distinguishing it from Short Range Devices is the lack of an internal power source in transponder. In the case of RFID systems energy is transmitted to transponders through a magnetic or electromagnetic field generated by a RFID reader antenna (Fig. 7). The same field is used for data transmission in both directions [19, 22].

In the moment of moving the transponder closer to the RFID reader antenna a process of charging an capacitor built in transponder [19] by field of the antenna occurs. After reaching a minimum voltage which enables to start of a microprocessor and process of communication begins. In HF RFID systems, compatible with ISO/IEC 14443 standard, Read/Write Device serves as a master device, whereas transponders are slave devices. It means that only RFID reader can start the communication process and send commands and transponders are only able to response to RFID reader commands. Transmission between them is held in half duplex mode according to command - response scheme (Fig. 8) [19, 23].

During transmission of a command the average value of energy transmitted to the transponder undergoes a slight decrease results from carrier wave amplitude's modulation. In case of transponder response transmission a RWD creates a constant carrier wave which enables power supply to transponder. The amplitude of wave reflected from transponder is modulated by transponder [19, 20]. It means that the most energy comes to the transponder before start of communication and during intervals between transmitted frames.

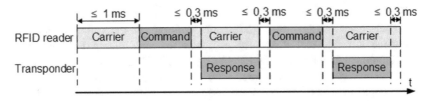

Fig. 8. Half duplex transmission in HF RFID system

The important module of the system is responsible for management of parking spaces in the city. RFID technology is employed to track occupancy of the places in real time. Parking places may be located along the streets or in car parks (parking lots). RFID transponders are located in such places to detect car presence by using RFID readers. This way, when a car takes a parking place, the cloud service gets an information that the place is occupied. Similarly, when the car leaves, the system immediately knows that the place is empty.

Currently, the most common solutions for parking management use infrared or ultrasonic sensors to detect whether a parking place is occupied or empty. By using RFID technology, one may reduce drawback of current solutions or introduce new features. For example, the system may charge the driver for parking without the need to take a ticket. The RFID reader in the vehicle will get access information from the identifier located at the entry to the car park. After acceptance, the car is guided to the nearest parking place.

Since the system manages all parking areas in the city, it can evenly distribute car park occupancy. By integrating traffic and car park management the system guides a car to the car park which is nearest the destination and which has a free place left. The system is able to predict the arrival time in advance, due to real-time traffic data.

Using the system application, the driver is able to reserve a parking place, so it is available upon the arrival. The selection is done either manually or automatically. In the first case, the user selects a car park, an address or a street. The automatic selection is done by indicating a parking place which is closest to the destination.

6 MQTT Communication with the Cloud

There are several protocol solutions for interfacing embedded devices with cloud services via 5G network (Fig. 1). For the purpose of traffic monitoring system presented here, we have chosen Message Queuing Telemetry Transport (MQTT) [25]. The protocol is particularly suited to low-bandwidth, high-latency or unreliable connections. MQTT, first developed in 1999, has gained its popularity with the growth of IoT (Internet of Things) solutions as a protocol for machine-to-machine (M2M) communication. It is especially useful for communication with embedded devices where network bandwidth and CPU limitations are considered like in the traffic monitoring system presented here. An alternative would be e.g. Constrained Application Protocol (CoAP) [26], however we found it rather oriented towards smart building solutions.

MQTT may be implemented in devices based on different CPUs, including ARM and AVR architectures [24]. In the laboratory prototype of the traffic system we have used popular Arduino and Raspberry platforms to create onboard devices for cars which communicate with the Azure cloud via MQTT using the publish/subscribe pattern. The communication is secured with token-based authentication.

The cloud-based services we have used in the traffic system are built around Microsoft Azure IoT Hub which implements MQTT protocol. It is worth to mention however, that other cloud vendors provide similar solutions. For example, there is IBM Watson IoT which offer comparable range of services and also handles MQTT protocol.

7   Azure Cloud Services

The traffic monitoring system uses a set of Microsoft Azure features to im-plement services for vehicles, like the guiding facility or the car park man-agement. The services have been implemented as a set of API functions for mobile and JavaScript clients, MQTT publishers for car embedded devices and a Web administration tool. Most of the software has been implemented in C# and JavaScript, with some Python scripts for machine learning.

The following features of Azure have been used:

- IoT Hub
- Cosmos DB database
- Location Based Services (LBS)
- Machine Learning Services.

IoT Hub implements and maintains MQTT connections between the system and vehicles. It may be seen as a bridge connecting the devices with higher-level control algorithms. This feature of Microsoft Azure allows not only to handle massive amount of data including data related to sensors/devices, but it also allows to maintain multiple connections and send data to all IoT devices within IoT Hub [27].

Data which is sent to the cloud is collected in Cosmos DB. It is a non-rela-tional (NoSQL) database which minor latency and high availability. Cosmos DB gives ability to store the data using various data models, such as key-value model, document data model or graph data model [28]. It also provides many APIs for programming pur-poses. In case of the traffic system we have used SQL queries to databases and JavaScript user procedures at the server side. Taking into consideration large amount of data processed in the system, scalability of the database engine is also important.

LBS services have been used to translate street addresses to map coordinates (geocoding) and to present maps to the driver and to operators of the system. Dynamic routing of a vehicle has also been made using LBS.

Azure Machine Learning is used to train the system to predict optimal traffic distribution. Data collected from cars and stored in Cosmos DB is selected to create the learning and testing sets. As the result, a model is created which takes into account such aspects as daytime (busy hours), weather or accidents to prepare an optimal route.

It is worth to mention that the machine learning process is specific to local conditions, so it should be adjusted for each city. Since the process is supported with the graphical tool called Machine Learning Studio, it is achieved with an easy-to-follow procedure. Figure 9 presents a screen from Microsoft Azure Machine Learning Studio showing a part of an experimental model. The experiment has been conducted using sample data gathered during laboratory tests stored in the .csv file. By using Machine Learning Studio, it was available to predict a road for specified vehicle. The data used for the tests include vehicle type, vehicle class, type of fuel, road availability, arrival time on each measurement point and simulated number of vehicles on each intersection. The laboratory tests involved also a number of cars simulated by a computer program.

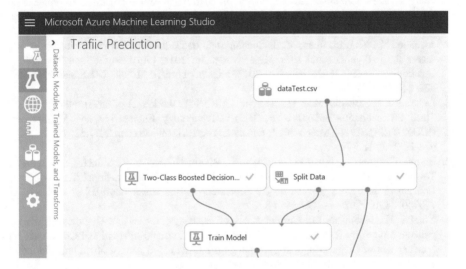

Fig. 9. A part of an experimental model in Microsoft Azure Machine Learning Studio presenting data prediction from uploaded dataTest.csv file

8 Conclusions

The presented concept has been tested using a laboratory prototype with a limited set of traffic data. This is of course not enough to make sure that the system will work flawlessly when it comes to a real scenario, especially for large-scale applications. However, due to high scalability of Azure cloud-based services and high performance of 5G network we can assume, that the system will work as expected.

The cloud-based solution makes it possible to introduce additional services to the system in the future. For example, one can process pictures taken by municipal traffic-monitoring cameras with some advanced vision algorithms to get information about traffic intensity and accidents on the roads. Azure Cognitive Services with Computer Vision API may be used for this purpose.

Acknowledgment. Results of Grant No. PBS1/A3/3/2012 from Polish National Centre for Research and Development as well as Statutory Activity of Rzeszow University of Technology were applied in this work. The work was developed by using equipment purchased in Operational Program Development of Eastern Poland 2007–2013, Priority Axis I Modern Economics, Activity I.3 Supporting Innovation under Grant No. POPW.01.03.00-18-012/09-00 as well as Program of Development of Podkarpacie Province of European Regional Development Fund under Grant No. UDA-RPPK.01.03.00-18-003/10-00.

References

1. Arasteh, H., Hosseinnezhad, V., Loia, V., Tommasetti, A., Troisi, O., Shafie-khah, M., Siano, P.: IoT-based smart cities: a survey. In: 2016 IEEE 16th International Conference on Environment and Electrical Engineering (EEEIC), 7–10 June 2016. https://doi.org/10.1109/eeeic.2016.7555867. ISBN: 978-1-5090-2320-2
2. Alshaflut, A., Thayananthan, V.: Estimating data traffic through software-defined multiple access for IoT applications over 5G networks. In: 2018 15th Learning and Technology Conference (L&T), 25–26 February 2018. https://doi.org/10.1109/lt.2018.8368486. ISBN: 978-1-5386-4817-9
3. Aidasani L.K., Bhadkamkar, H., Kashyap, A.K.: IoT: the kernel of smart cities. In: 2017 Third International Conference on Science Technology Engineering and Management (ICONSTEM), 23–24 March 2017. https://doi.org/10.1109/iconstem.2017.8261248. ISBN: 978-1-5090-4855-7
4. Barreto Campos, L., Eduardo Cugnasca, C., Riyuiti Hirakawa, A., Sidnei, C. Martini, J.: Towards an IoT-based system for Smart City. In: 2016 IEEE International Symposium on Consumer Electronics (ISCE), 28–30 September 2016. https://doi.org/10.1109/isce.2016.7797405. ISBN: 978-1-5090-1549-8
5. Yutaka, M., Nobutaka, O.: Current status of smart systems and case studies of privacy protection platform for smart city in Japan. In: 2015 Portland International Conference on Management of Engineering and Technology (PICMET), 2–6 August 2015. https://doi.org/10.1109/picmet.2015.7273158. ISBN: 978-1-8908-4331-1
6. Jabbarpour, M.R., Nabaei, A., Zarrabi, H.: Intelligent guardrails: an IoT application for vehicle traffic congestion reduction in smart city. In: 2016 IEEE International Conference on Internet of Things (iThings) and IEEE Green Computing and Communications (GreenCom) and IEEE Cyber, Physical and Social Computing (CPSCom) and IEEE Smart Data (SmartData), 15–18 December 2016. https://doi.org/10.1109/ithings-greencom-cpscom-smartdata.2016.29. ISBN: 978-1-5090-5880-8
7. Javaid, S., Sufian, A., Pervaiz, S., Tanveer, M.: Smart traffic management system using Internet of Things. In: 2018 20th International Conference on Advanced Communication Technology (ICACT), 11–14 February 2018. https://doi.org/10.23919/icact.2018.8323770. ISBN: 979-11-88428-01-4
8. Wong, S.F., Mak, H.C., Ku, C.H., Ho, W.I.: Developing advanced traffic violation detection system with RFID technology for smart city. In: 2017 IEEE International Conference on Industrial Engineering and Engineering Management (IEEM), 10–13 December 2017. https://doi.org/10.1109/ieem.2017.8289907. ISBN: 978-1-5386-0948-4
9. Moran, O., Gilmore, R., Ordóñez-Hurtado, R., Shorten, R.: Hybrid urban navigation for smart cities. In: 2017 IEEE 20th International Conference on Intelligent Transportation Systems (ITSC), 16–19 October 2017. https://doi.org/10.1109/itsc.2017.8317858. ISBN: 978-1-5386-1526-3

10. Pedraza, C., Vega, F., Manana, G.: PCIV, an RFID-based platform for intelligent vehicle monitoring. In: IEEE Intelligent Transportation Systems Magazine, vol. 10, no. 2, pp. 28–35, Summer 2018. https://doi.org/10.1109/mits.2018.2806641. ISSN: 1939-1390
11. AlBraheem, L., Al-Abdulkarim, A., Al-Dosari, A., Al-Abdulkarim, L., Al-Khudair, R., Al-Jasser, W., Al-Angari, W.: Smart city project using proximity marketing technology. In: 2017 Intelligent Systems Conference (IntelliSys), 7–8 September 2017. https://doi.org/10.1109/intellisys.2017.8324288. ISBN: 978-1-5090-6435-9
12. Farhad, S.M., Alahi, I., Islam, M.: Internet of Things based free parking space management system. In: 2017 International Conference on Cloud Computing Research and Innovation (ICCCRI), 11–12 April 2017. https://doi.org/10.1109/icccri.2017.8. ISBN: 978-1-5386-1075-6
13. Joshi, Y., Gharate, P., Ahire, C., Alai, N., Sonavane, S.: Smart parking management system using RFID and OCR. In: 2015 International Conference on Energy Systems and Applications, 30 October–1 November 2015. https://doi.org/10.1109/icesa.2015.7503445. ISBN: 978-1-4673-6817-9
14. Abdulkader, O., Bamhdi, A.M., Thayananthan, V., Jambi, K., Alrasheedi, M.: A novel and secure smart parking management system (SPMS) based on integration of WSN, RFID, and IoT. In: 2018 15th Learning and Technology Conference (L&T), 25–26 February 2018. https://doi.org/10.1109/lt.2018.8368492. ISBN: 978-1-5386-4817-9
15. Ming-Shen, J., Shu Hui, H.: Location aware public/personal diversity of information services based on embedded RFID platform. In: 11th International Conference on Advanced Communication Technology, ICACT 2009, 15–18 February 2009. ISBN: 978-89-5519-138-7
16. Prateek, J.: Artificial Intelligence with Python, pp. 14–16. Packt Publishing, Birmingham (2017)
17. Jankowski-Mihułowicz, P., Węglarski, M., Pitera, G., Kawalec, D., Lichoń, W.: Development board of the autonomous semi-passive RFID transponder. Bull. Pol. Acad. Sci. Tech. Sci. 64(3), 647–654 (2016). https://doi.org/10.1515/bpasts-2016-0073
18. Jankowski-Mihułowicz, P., Węglarski M.: Definition, characteristics and determining parameters of antennas in terms of synthesizing the interrogation zone. In: Crepaldi, P.C., Pimenta, T.C. (eds.) RFID Systems, Radio Frequency Identification, Chap. 5, pp. 65–119. INTECH, 29 November 2017. https://doi.org/10.5772/intechopen.71378. ISBN 978-953-51-3630-9
19. Finkenzeller, K.: RFID Handbook—Fundamentals and Applications in Contactless Smart Cards and Identification. Wiley, New York (2003)
20. ISO/IEC 14443-1: Identification cards—Contactless integrated circuit cards—proximity cards—part 1: physical characteristics (2016). https://www.iso.org/standard/70170.html
21. ISO/IEC 14443-2: Identification cards—contactless integrated circuit cards—proximity cards—part 2: radio frequency power and signal interface (2010). https://www.iso.org/standard/50941.html
22. ISO/IEC 14443-3: Identification cards—contactless integrated circuit cards—proximity cards—part 3: initialization and anticollision (2011). https://www.iso.org/standard/50942.html
23. ISO/IEC 14443-4: Identification cards—contactless integrated circuit cards—proximity cards—part 4: transmission protocol (2016). https://www.iso.org/standard/70172.html
24. Hillar, G.C.: MQTT Essentials: A Lightweight IoT Protocol. Packt Publishing, Birmingham (2017)
25. MQ Telemetry Transport Protocol. http://mqtt.org
26. The Constrained Application Protocol (CoAP). https://tools.ietf.org/html/rfc7252
27. Crump, M., Luijbregts, B.: The Developer's Guide to Azure, 2nd edn. Microsoft Press, Redmond (2018)
28. Guay Paz, R.J.: Microsoft Azure Cosmos DB Revealed. Apress, New York (2018)

Monitoring and Maintenance of Telecommunication Systems: Challenges and Research Perspectives

Lakmal Silva[1], Michael Unterkalmsteiner[2], and Krzysztof Wnuk[2(✉)]

[1] Ericsson AB, Karlskrona, Sweden
ruwan.lakmal.silva@ericsson.com
[2] Software Engineering Research Lab, Blekinge Institute of Technology,
Karlskrona, Sweden
{michael.unterkalmsteiner,krw}@bth.se

Abstract. In this paper, we present challenges associated with monitoring and maintaining a large telecom system at Ericsson that was developed with high degree of component reuse. The system constitutes of multiple services, composed of both legacy and modern systems that are constantly changing and need to be adapted to changing business needs. The paper is based on firsthand experience from architecting, developing and maintaining such a system, pointing out current challenges and potential avenues for future research that might contribute to addressing them.

Keywords: Legacy system evolution · Virtualization
Telecommunication services

1 Introduction

The Telecom industry has undergone a substantial transformation in the last few years, working towards fifth generation (5G) networks [1]. Ericsson is no exception and has been experiencing an increased pressure to lower operational costs and quickly respond to market changes. Therefore, some development units within Ericsson made a strategic decision in 2013 to increase the reusability of components from the existing portfolio, preserving the investment of many years of tuning and debugging [2]. With this directive, a new system was built reusing four products from the portfolio. This new system is designed to manage the life cycle of virtual resources, such as virtual machines (VMs) and networks. Reuse certainly sped up the development, reduced time to market along with the added benefits of quality, reused software architecture, infrastructure, and domain knowledge [3]. However, at the same time, we encountered challenges related to the interoperability of legacy products, integrating reuse practices into the development process, and the deployment of new services into cloud environments. We provide first-hand insight into these challenges and point out avenues for future research that may address these challenges.

© Springer Nature Switzerland AG 2019
P. Kosiuczenko and Z. Zieliński (Eds.): KKIO 2018, AISC 830, pp. 166–172, 2019.
https://doi.org/10.1007/978-3-319-99617-2_11

The remainder of the paper is structured as follows. Section 2 briefly reports on related work. Section 3 introduces the industrial context of the challenges which are further expanded in Sect. 4. Section 5 concludes the paper.

2   Related Work

Khadka et al. studied legacy to service-oriented architecture (SOA) evolution and concluded that reverse engineering is the most common technique for legacy system understanding [4] and wrapping is the most common implementation technique. Almonaies et al. [2] surveyed the approaches to moving legacy systems to the SOA environment with the help of redevelopment, replacement, wrapping and migration strategies. Apart from SaaS migration, cloud migration is an emerging research area with various aspects of legacy-to-cloud migration [5]. Jamshidi et al. provide an interesting comparison of SOA and cloud migration in the drivers, provisioning, design principles and crosscutting concerns perspectives. Toffetti et al. [6] propose a new architecture that enables scalable and self-managing applications in the cloud while Pahl et al. surveyed cloud container technologies and architectures [7] that offer lightweight virtualization.

Balalaie et al. described the incremental migration and architectural refactoring of a commercial mobile back-end as a service to microservices architecture with the help of "Development" and Operations" DevOps [8]. Finally, Varghese and Buyya discussed new trends and research directions in next generation cloud computing [9] while Xavier and Kantarci surveyed challenges and opportunities for communication and network enablers for cloud-based services [10].

3   Industrial Context

The system we discuss in this paper was designed to manage the life-cycle of virtual resources, such as virtual machines, virtual networks in multi-vendor virtual infrastructures such as open stack and vSphere, as well as to orchestrate virtual network functions (VNFs).

The software components are reused when composing complex subsystems. Due to the monolithic nature of the subsystems, it was not possible to extract only the required functionality needed for the new system. As a side effect, the actual percentage of utilization of the functionality from each embedded subsystem was low (between 20-30%). On the other hand, there is redundant functionality among different subsystems. For instance, each subsystem has its own mechanisms for authentication and authorization. These are core functionalities within each subsystem, so it is difficult to bypass them without significant design changes. Presence of these functionality contributes to extra processing overhead, as well as adding complexity to the operation and maintenance tasks such as changing of system passwords.

The network setup of each subsystem became a challenge when wrapping them into VMs as each has its own flavor of setting up the network. Although Ericsson has set guidelines and requirements for setting up the networks, the actual implementation of each product has its own flavor when it comes to detailed setup.

The functionality of the new system had to be mapped to fit into the existing architecture, rather than designing a system that best suits the business requirements. This led to an architectural degradation and consequences in the systems runtime characteristics which are explained next.

3.1 System Architecture

As illustrated in Fig. 1, the reused subsystems expose different inbound interfaces. The Representational state transfer (REST) interface in VM1 is the main entry point into the new system. This interface can be used to build customer specific services without having to change the core system. The main service provided by VM1 is to slice the order into fine grained requests to be orchestrated towards VM3, and to store the necessary data in the database located in VM2. Since these subsystems have been developed independently in different timelines by different development units, the protocols and technologies used vary (REST in VM1, Structured Query Language (SQL) in VM2 and Simple Object Access Protocol (SOAP) in VM3). When building a solution reusing these systems, there is a significant amount of overhead due to protocol transformations between the involved services.

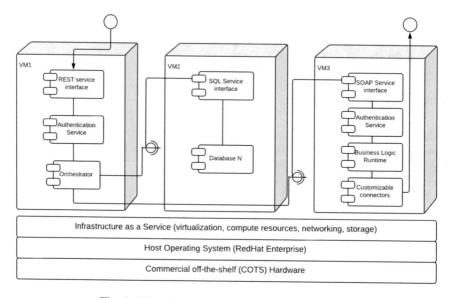

Fig. 1. Virtual resource manager system architecture

Similarly, the subsystems have their own ways of managing configuration. Because of this, the administrators of the system won't experience a uniform configuration management. The administrator is also expected to understand the dependencies between configurations of multiple subsystems. For instance, if a password is changed in VM1, the administrator is expected to have prior knowledge that that the password in VM2 and VM3 should also be changed. There is no centralized configuration management which distributes the configuration throughout the involved subsystems.

Different subsystems also come with a variety of databases. Their maintenance and configuration require varying techniques, forcing the administrators to acquire competency for multiple database management systems. When issues arise at customer sites, the Ericsson development unit needs to get involved, supporting the customer sites' operation and maintenance staff. This consumes time and resources that has been allocated for feature development of future releases.

3.2 Runtime System Characteristic

The reused subsystems were originally not developed to be run on virtualized or cloud platforms which is a strict requirement for the new system. Similar to the wrapping strategies described by Almonaies et al. [2], our approach was to wrap the legacy software into virtual machines. The resulting deployment architecture in a cloud stack is shown in Fig. 1. Some of the issues detailed next are the results of trying to fit legacy subsystems into modern virtualized and cloud environments.

The installation procedures of the involved subsystems vary due to their independent development by separate design units at Ericsson. These variants can be managed by using configuration and automation tools, but it is time consuming to develop a uniform framework for the new system. To add another dimension to the complexity, the subsystems are developed in different countries, driving significant delays in communication due differences in time zones.

Monitoring and recovery from process failures among the reused subsystems is not uniform. Few of the subsystems provide monitoring agents which automatically recover process failures along with sending failure and recovery notifications to the monitoring systems, whereas most subsystems provide no process recovery at all.

The complexity of the cloud stack creates problems in the areas of performance tuning and troubleshooting. Different vendors provide different layers of the cloud stack and different teams configure these layers. The performance of the systems relies on the efficiency of different layers in the cloud stack, all the way down to the hardware. The system can misbehave due to issues in any of layer of the stack, forcing support teams from each layer to get involved to resolve issues; this process can be very time consuming.

4 Challenges and Research Directions

Building new services/systems by reusing existing products may seem like a good approach to go to market faster and to tap into new domains with a lower cost. However, this approach comes at a cost in the long run if these systems do not evolve along technological advancements. Table 1 identifies the key challenges, specifically when reusing large legacy systems. Next, we discuss these challenges, pointing out potential avenues for further research.

Table 1. Challenges related to legacy system reuse

Id	Challenge	Description and specifics
Ch1	Interoperability of legacy and new services	The reused subsystems are a mixture of both legacy systems and relatively modern systems which creates interoperability issues
Ch2	Development process for system reuse	Even though modern development processes such as DevOps have been adopted in the development phase, it is not efficient due to the characteristics of the reused subsystems
Ch3	Deployment and orchestration of services	Issues related installation, monitoring, scaling of the system in virtualized and cloud environments

Ch1: Existing systems are adapted to provide services that were not intended to be supported by the original system, which results in issues being discovered very late in the development cycle and at customer sites. Another issue with the current architecture is the overhead involved in the overall system, as different subsystems are required to perform protocol transformations.

Ch1.1: The system is required to comply with certain standards such as ETSI GS NFV-MANO demanded by customers. This adds to the complexity of the system architecture, as standardization was not regarded important when the legacy systems were initially designed.

Research Directions: Previous research has illustrated strategies for migrating monolithic to microservices-based architectures that allow for flexible reuse [8, 11]. The environment at Ericsson requires however adaptations to these strategies, since development is globally distributed, and the product needs to be standard compliant.

Ch2: The development unit at Ericsson responsible for the system has already adopted DevOps practices such as Continuous Integration and Deployment (CI/CD). However, the subsystems are so complex and large such that the lengthy CI/CD cycle renders development inefficient.

Ch2.1: The development teams are distributed in at least five geographically dispersed locations. To facilitate management, the teams were organized as feature teams, that is, when a feature was handed over to a design team, they were responsible for the end-to-end development of the feature, which involved adaptations to a few complex subsystems. Due to the complexity of each subsystem, there are no subsystem experts in teams, resulting in misuse of the architecture of the subsystems and therefore contributing to the degradation of the overall system architecture.

Research Directions: DevOps practices need to be improved to facilitate the growing complexity in platforms and IT operations to prevent DevOps practices becoming an overhead in the development process. As others have observed [12], insufficient communication, company culture, legal constraints and heterogeneous environments can prevent successful DevOps practices adoption. Similar causes can be observed at Ericsson and dedicated solutions need to be investigated.

Ch3: The reused subsystems were not designed to be deployed into virtual environments, and the developed wrappers and frameworks soon became complex and are hard to manage.

Ch3.1: The demands on the operation and maintenance staff rose to the same degree as the complexity of the overall system increased. This contributes to increased system maintenance and troubleshooting time, due to inhomogeneous and scattered information in subsystems.

Ch3.2: When the system is deployed into cloud environments, it has also to be continuously monitored and managed [6]. The basis for efficient management of cloud-based services is to slice the monolithic applications into microservices [6]. The current strategy to wrap monolithic subsystems into a single VM makes it difficult to extract subsystems into their own services and be managed independently.

Research Directions:
There is a need to provide support for the operation and maintenance staff to interpret and analyze the data collected in production. While this data is useful for troubleshooting, it requires expertise that is not always available and costly. One possibility is to combine big data analytics and machine learning to support operation staff in decision making by filtering and identifying patterns in data [13, 14].

5   Conclusions and Future Work

The demand for new and more features, at a faster pace and at a lower cost, has affected all high-tech industries. This requires that well established players embrace this change and identify strategies that allow them to reuse existing technologies, adapt them to new requirements, and can provide thereby new services on time, quality and cost. In this paper, we have illustrated some of the challenges related to this transition, seen through the lens of an experienced architect at Ericsson. Future work is targeted at providing a systematic, multi-perspective investigation to prioritize the challenges and guide solution development.

References

1. Chen, S., Zhao, J.: The requirements, challenges, and technologies for 5G of terrestrial mobile telecommunication. IEEE Commun. Mag. **52**(5), 36–43 (2014)
2. Almonaies, A.A., Cordy, J.R., Dean, T.R.: Legacy system evolution towards service-oriented architecture. In: International Workshop on SOA Migration and Evolution (SOAME), Madrid. IEEE (2010)
3. Mohagheghi, P., Conradi, R.: An empirical investigation of software reuse benefits in a large telecom product. ACM Trans. Softw. Eng. Methodol. **17**(13), 1–31 (2008)
4. Khadka, R., Saeidi, A., Idu, A., Hage, J., Jansen, S.: Legacy to SOA evolution: a systematic literature review. In: Migrating Legacy Applications: Challenges in Service Oriented Architecture and Cloud Computing Environments, pp. 40–70. IGI Global (2013)

5. Jamshidi, P., Ahmad, A., Pahl, C.: Cloud migration research: a systematic review. IEEE Trans. Cloud Comput. **1**(2), 142–157 (2013)
6. Toffetti, G., Brunner, S., Blöchlinger, M., Spillner, J., Bohnert, T.M.: Self-managing cloud-native applications: design, implementation, and experience. Future Gen. Comput. Syst. **72**, 165–179 (2017)
7. Pahl, C., Brogi, A., Soldani, J., Jamshidi, P.: Cloud container technologies: a state-of-the-art review. IEEE Trans. Cloud Comput. 1–14 (2017)
8. Balalaie, A., Heydarnoori, A., Jamshidi, P.: Microservices architecture enables DevOps: migration to a cloud-native architecture. IEEE Softw. **33**(3), 42–52 (2016)
9. Varghese, B., Buyya, R.: Next generation cloud computing: New trends and research directions. Future Gener. Comput. Syst. **79**, 849–861 (2018)
10. Xavier, G.P., Kantarci, B.: A survey on the communication and network enablers for cloud-based services: state of the art, challenges, and opportunities. Ann. Tel. **73**(3), 169–192 (2018)
11. Dragoni, N., Dustdar, S., Larsen, S.T., Mazzara, M.: Microservices: migration of a mission critical system (2017)
12. Riungu-Kalliosaari, L., Mäkinen, S., Lwakatare, L.E., Tiihonen, J., Männistö, T.: DevOps adoption benefits and challenges in practice: a case study. In: Abrahamsson, P., Jedlitschka, A., Nguyen, Duc A., Felderer, M., Amasaki, S., Mikkonen, T. (eds.) Product-Focused Software Process Improvement, pp. 590–597. Springer, Cham (2016)
13. Zaman, F., Hogan, G., Meer, S.V.D., Keeney, J., Robitzsch, S., Muntean, G.M.: A recommender system architecture for predictive telecom network management. IEEE Commun. Mag. **53**(1), 286–293 (2015)
14. Parwez, M.S., Rawat, D.B., Garuba, M.: Big data analytics for user-activity analysis and user-anomaly detection in mobile wireless network. IEEE Trans. Ind. Inf. **13**(4), 2058–2065 (2017)

Empirical Software Engineering

Software Metrics for Similarity Determination of Complex Software Systems

Andrzej Stasiak$^{(\boxtimes)}$ ⓘ, Jan Chudzikiewicz ⓘ,
and Zbigniew Zieliński ⓘ

Faculty of Cybernetics, Military University of Technology,
gen. Urbanowicza 2, 00-908 Warsaw, Poland
{andrzej.stasiak, jan.chudzikiewicz,
zbigniew.zielinski}@wat.edu.pl

Abstract. In the paper a methodology for similarity determination of complex software systems is proposed. The methodology introduce definition of software systems similarity metrics and ways to its determination, as well methods and tools for measuring similarity with complete environment. The essence of the proposed methodology is to divide of software of systems to disjoint groups on the base of proposed structural similarity metrics to significantly reduce of computational complexity of necessary comparisons of source code. Presented environment for similarity determination of software systems using .NET technology was utilized for preparing of opinions in legal proceedings for copyright violation.

Keywords: Software metrics · Measuring similarity
Structural similarity metrics

1 Introduction

The paper concerns the issue of examining the similarity of source code of IT system software. The essential motivation for conducting the research presented in this paper was participation in the analyses of IT system software similarity for the purposes of court proceedings concerning copyright infringement. In these trials, the goal was most commonly to determine the degree of source code similarity of the examined system software. Due to the complexity of the analyzed systems' software (the source code included as many as several thousand files), determination of software similarity of the systems was not an easy task, primarily due to calculation complexity. Files containing the software of the analyzed systems differed in names, so a search for source code matches would require testing the similarity of every possible pair of files, where one element would be a file of system X, and the other - a file of system Y. Such a procedure would be extremely time-consuming, as it would require performing millions of file comparisons. One method of reducing the calculation complexity of comparative tests is to divide software into certain classes and to test similarity only within such file classes.

© Springer Nature Switzerland AG 2019
P. Kosiuczenko and Z. Zieliński (Eds.): KKIO 2018, AISC 830, pp. 175–191, 2019.
https://doi.org/10.1007/978-3-319-99617-2_12

Similar issues with similarity detection can also, in a general manner, concern document similarity testing [1, 2], as well as the determination of similarity and assessment of evolutionary changes of IT system source code [3, 4]. Similar issues also emerge in text plagiarism detection [5, 6]. An overview of related papers and problems is presented in Sect. 5.

For the general issue of document similarity testing, it is assumed that the basic element is the document, which constitutes a single file. The file contains text in the form of strings of characters, forming expressions in the given programming language. The concept of the method can be reduced to searching documents for the same series of programming language instructions (so called idioms – directly present in the files where the source code is stored). Patterns defined as sequences of instructions common for both systems are identified and such parameters as their length and frequency of appearances relative to code volume (expressed usually as the number of lines of code) are measured. In this sense, the proposed method enables identification and measurement of patterns present in both systems.

Section 1 presents a formal definition of software similarity and specifies its measures – the global measure of software similarity, familiar from [4], and a measure of structural similarity of software code file groups, proposed by the authors. Section 2 describes the approach to software similarity testing and proposes performing measurements using source code description structures. Section 3 describes a practical Methodology of Software Similarity Study (M3S) and presents the developed and applied Software Similarity Study Tool (3ST). Section 4 presents sample results for software testing (specific to the .NET platform), obtained in the 3ST environment. Section 5 presents our results in comparison with related studies. Section 6 contains end comments and indicates the directions for further studies.

2 Similarity of Complex Software Systems

Conventional wisdom in the field of software engineering, the use of metrics mainly applies to controlling software quality, efficiency and software production processes, or its testing. Unlike the above, the primary goal of this study was quantitative assessment of system similarity, which involved determining system software similarity measures and their values.

2.1 Global Software Similarity Measure

This paper assumes that system software similarity testing will concern a set of IT systems $S = \{S_1, S_2, \ldots, S_z\}$ relative to a certain reference system X. It bears noting that, for example: subsequent s_j may be newer versions of s_{j-1} systems. The issue as defined in this way is discussed in [4]. We will thus compare the software of a pair of systems X and Y, where Y refers to the fixed element of the set S, which is $Y \in \{S_1, S_2, \ldots S_z\}$.

Further, we assume that systems X and Y are made up of elements based on which their software is built, i.e. files with source code, constituting the end result of the creative work of programmers[1], where: $X = \{x_1, x_2, \ldots x_m\}$; $Y = \{y_1, y_2, \ldots y_n\}$; and $(1 \leq m \leq M, 1 \leq n \leq N)$. Further, based on [4], we define the general measure of similarity SIM for systems X and Y as:

$$SIM(X, Y) \equiv \frac{\left|\{x_i | (x_i, y_j) \in R_s\}\right| + \left|\{y_j | (x_i, y_j) \in R_s\}\right|}{|X| + |Y|} \tag{1}$$

where R_S (Fig. 1) is a general relation of correspondence between systems x_i and y_j, which may for example concern lines of code [7], but also other characteristics, e.g. architecture elements expressed in the structure of classes and components.

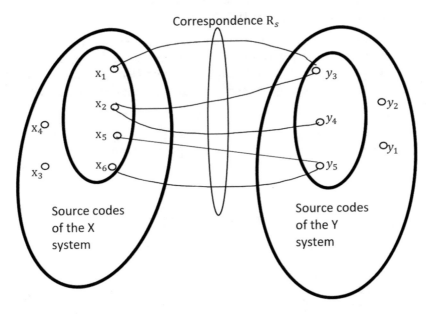

Fig. 1. Correspondence of elements R_s.

2.2 Measure of Structural Similarity of Software Code File Groups

Given that the systems to be analyzed can comprise many thousands of source files, determining this relation for all file pairs poses a complex calculation issue. In order to constrain it, we propose reducing the se of the task by taking into account the similarity within software layers. Most commonly, software of complex systems encompasses at least three of its layers, which include: the presentation layer, the business logic layer, and the data access layer. In the proposed method, as a simplification, layers have their corresponding source code file groups – GSC^g. Because currently, multiple programming languages are usually used, for example for building the presentation layer, then aside from the implementation language, "inserts" in HTML and/or JavaScript are

---

[1] Object code is ignored.

usually used, and consequently the concept of layer was replaced with the concept of source code file group GSC^G, where G means the number of types of source code file groups.

Example 1. During the study, similarity of systems implemented using the .NET platform was analysed, therefore the following types of source code file groups were identified: $GSC^g \in \{ASCX, ASPX, HTML, ASCX\_CS, ASPX\_CS, CS\}$. As indicated by GSC^g analysis, the presentation layer in the example is created by:

- definitions of interface page elements - .aspx files,
- handling interface pages - .aspx.cs files,
- handling interface own controls - .ascx and .ascx.cs files,
- application interface elements in HTML - .htm files,

and the business logic layer and data access were implemented in C# (CS).

For this reason, we propose a re-definition of the general relation of correspondence between elements of systems x_i *and* y_j as a relation concerning groups of code files $R_s^{GSC^g}$ (Fig. 2) and to introduce the concept of structural similarity of software groups SIMstr.

$$SIMstr\left(X^{GSC^g}, Y^{GSC^g}\right) \equiv \frac{\left|\{x_i | (x_i, y_j) \in R_s^{GSC^g}\}\right| + \left|\{y_j | (x_i, y_j) \in R_s^{GSC^g}\}\right|}{|X^{GSC^g}| + |Y^{GSC^g}|} \qquad (2)$$

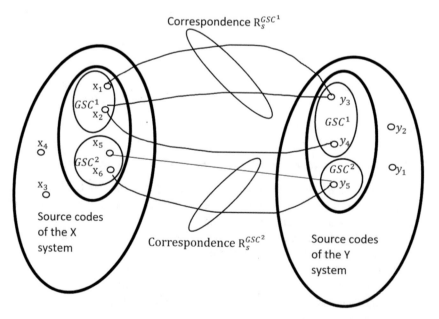

Fig. 2. Correspondence of $R_s^{GSC^g}$ elements within groups of source code

2.3 Structural Similarity Metrics

Let us look at the codes of systems X and Y through the grammar of the programming languages in which the systems were implemented and for which the following syntax-controlled analyzers were developed: $LA^g \in \{$ASCX, ASPX, HTML, ASCX_CS, ASPX_CS, CS$\}$. It was assumed that for each software group GSC^g a dedicated LA^g analyser will be developed.

As a simplification, we can assume that the grammar of the implementation language of each source code file group GSC^g determines the language structure type, which we are going to call structure types $Type_t^g$, where $t = 1..T(GSC^g)$, as well as their characteristics, which we are going to call structure elements $Elem_e^t$, where $e = 1..E$.

Example 2. The quantities of the structure types "t" proposed in the paper, for individual source code file groups "g" – .NET platform, which formed the environment for implementing systems X and Y, are shown in Table 1.

Table 1. Quantities of software code file groups.

	ASCX	ASPX	HTML	ASCX_CS	ASPX_CS	CS
$g =$	1	2	3	4	5	6
$T(GSC^g) =$	12	15	11	14	14	14

While structure types $Type_t^g$ are shown in the Table 2.

It is worth noting that as in flowcharts, structure types denote block types (e.g. edge block, I/O block, calculation, decision, subroutine call, fragment or comment block), while for programming languages, they denote the programming language structure types corresponding to these blocks.

For the purpose of testing software similarity, in the proposed approach we create descriptions of source code. These descriptions constitute an ordered set of instances of language structure types and elements, and are the result of work of lexical analyzers LA^g. We refer to these descriptions as tons (similar to [7]). Tokens are created for systems X and Y. Thus, for system X, we can assume that a token constitutes a specific source code fragment of file x_i belonging to code group GSC^g, described by the pair $\langle Type_{t'}^g(x_i), \{Elem_{e'}^{t'}(x_i)\} \rangle$, while for system $Y - \langle Type_{t''}^g(y_j), \{Elem_{e''}^{t''}(y_j)\} \rangle$, where: $g = 1..G; t' = 1..T; t'' = 1..T; e' = 1..E; e'' = 1..E$.

The xsd structure, based on which the token is created, is shown in Fig. 3.

As has been noted before, software similarity of systems X and Y should be tested only in relation to their parts x_i and y_j that belong to the same source code file groups GSC^g, i.e. which use the same structure types $Type_t^g$, shown in Table 2.

It is worth noting that tokens create a new dimension of software structure description – a new view, where the structures used do not have to be directly assigned

Table 2. List of structure types of source code file groups of systems X and Y.

ASCX	ASPX	HTML	ASCX_CS	ASPX_CS	CS
Attribute	Attribute	Attribute	Class	Class	Class
AttributeValue	AttributeValue	AttributeValue	Comment	Comment	Comment
Comment	Comment	Comment	Indexer	Indexer	Indexer
Control	Control	CssClass	Instruction	Instruction	Instruction
CssClass	CssClass	CssPropety	KeyWord	KeyWord	KeyWord
CssPropety	CssPropety	DocumentType	Method	Method	Method
Directive	Directive	Java Script	Modifier	Modifier	Modifier
EmbeddedCode	DocumentType	PropertyValueCss	Name	Name	Name
JavaScript	EmbeddedCode	Tag	Namespace	Namespace	Namespace
PropertyValueCss	JavaScript	TagOptions	Operators	Operators	Operators
Tag	PropertyValueCss	Value	Property	Property	Property
Value	Selector		Text	Text	Text
	Tag		Type	Type	Type
	TagOptions		Variable	Variable	Variable
	Value				

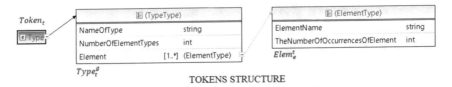

TOKENS STRUCTURE

Fig. 3. The xsd structure model for all types of tokens

to source code files, but only to the given software group[2] (e.g. its layer). Additionally, in the sense of structural similarity of software groups, only these subsets of X and Y should be compared which belong to the same source code file groups GSC^g. Based on the assumptions specified in point 1.1, we know that system X creates m source files, which are divided into separable subsets determined by groups GSC^g, which means that the total number of source files belong to all groups = m, and for system Y it equals n. We can therefore assume that each source code file $x_m \in X; y_n \in Y;$ is described by a sequence of tokens. For system X:

$x_m \xrightarrow{\text{is described as}} \langle Token_1(x_m), Token_2(x_m), \ldots, Token_T(x_m) \rangle,$ and for system Y
$y_n \xrightarrow{\text{is described as}} \langle Token_1(y_n), Token_2(y_n), \ldots, Token_T(y_n) \rangle.$

Because, in addition to type and element values, each token maintains information about their quantities (NumberOfElementTypes, TheNumberOfOccurencesOfElement), then on its basis a signature is created, which will be used to assess software similarity.

The signature is a vector of numbers, determined on the basis of source code software tokens (Fig. 3) from a file of identifier x_m or y_n, belonging to group GSC^g:

$$S^{GSC^g}(x_m) = \langle L_1(x_m), L_2(x_m), \ldots, L_{T(GSC^g)}(x_m) \rangle,$$

where: $L_t(x_m)$ – TheNumberOfOccurencesOfElement - number of syntax elements $Elem_t^{GSC^g}(x_i)$ of the software structure t, specified for software group GSC^g, and $t = 1..T(GSC^g)$.

Group similarity for systems $S(X)$ and $S(Y)$ is determined using the following relation:

$$\widehat{S^{G_i}}(X) = \langle L_1^{GSC^g}(X), L_2^{GSC^g}(X), \ldots, L_T^{GSC^g}(X) \rangle,$$

and:

$$\widehat{S^{G_i}}(Y) = \langle L_1^{GSC^g}(Y), L_2^{GSC^g}(Y), \ldots, L_2^{GSC^g}(Y) \rangle,$$

where: $L_i(X)$, $L_i(Y)$ are calculated as values of sum $L_i(x)$ for each file x with codes of type (GSC^g) for systems $S(X)$ and $S(Y)$, respectively.

---

[2] This is a very important result for the method, because "artificial" multiplication of source files will be detected as a result of the functioning of the proposed method.

$$SIMstr^{GSC^g}, \; w_t^{GSC^g}$$

Global similarity index SIMstr for the given software group SGC^g is expressed by Eqs. (3) and (4):

$$SIMstr^{GSC^g} = \frac{\sum_{t=1}^{t=T(GSC^g)} w_t^{GSC^g}}{T(GSC^g)}, \; \text{where} \tag{3}$$

$$w_t^{GSC^g} = 1 - \frac{\left| L_t(X^{GSC^g}) - L_t(Y^{GSC^g}) \right|}{L_t(X^{GSC^g}) + L_t(Y^{GSC^g})}. \tag{4}$$

Because structure types NameOfType - $Type_t^g(x_m)$ and $Type_t^g(y_n)$ result from the language grammar, their structure type dictionaries (specified in Table 2) should be identical for the analysed source files of systems X and Y (Fig. 4).

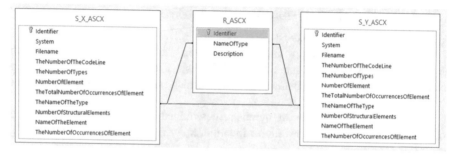

Fig. 4. Data model fragment for a group of ASCX code

Element values ElementName - $Elem_{e'}^{t'}(x_i)$ and $Elem_{e''}^{t''}(y_i)$ correspond to the "creative work of programmers" and in the case of original source code should take on unique values. This uniqueness can easily be disrupted by introducing "minor changes" - usually a constant prefix or suffix. Such disruptions can be detected by analysing the value dictionary (ElementName for the given NameOfType) or by analysing it using the Levenshtein metric [8] (detecting so called synonyms).

Thanks to such an approach, structural similarity **SIMstr** is determined on the basis of quantitative measures (specified by NumberOfElementTypes TheNumberOfOccurencesElement), and subsequently similarity is determined by comparing the values of types and elements (i.e. ElementName and NameOfType).

It is proposed to classify metrics as follows:

- GLO - global - related to the similarity of systems X and Y;
- GRU - group - related to separable file groups GSC^g within systems X and Y;
- TYP - type - i.e. related to the specific software structure type "t" - $Type_t^g$ within a specific group of source code files "g";
- LOC - local - i.e. related to files within the given system;

The metrics proposed in this paper are classified as one of these four categories {GLO, GRU, TYP, LOC} and apply to the following characteristics of software, determined on the basis of data recorded in the tokens.

- number of lines of source code;
- number of types of elements;
- number of elements;
- e total number of occurrences of elements;
- and derived characteristics, i.e.:
- number of files;
- global similarity index SIMstr;
- SLINE, as the number of identical lines of code in the files of the analysed systems.

3 An Approach to Similarity Measuring

For each file type group GSC^g that is used both in system X and Y, a separate lexical analyser was created - LA^g. As a lexical analyser operates, a description of the source code subject to the analysis is created in the form of an ordered set of tokens, $\forall(m,t), \exists(g), LA^g(x_m) \xrightarrow{create} \langle Token_1^g(x_m), \ldots, Token_t^g(x_m) \rangle$ for system X and $\forall(n,t),$ $\exists(g), LA^g(y_n) \xrightarrow{create} \langle Token_1^g(y_n), \ldots, Token_t^g(y_n) \rangle$ for system Y. Each source code description is a measurement result and defines a constant set of characteristics divided into two groups. Basic and additional characteristics (Fig. 5).

Basic characteristics determine the identity of a source code file (enable its identification) and indicate its complexity:

- file name <FN> and number of lines of source code <NLSC>;

as well as describe structural similarity by specifying:

- number of types of elements <NTE>, number of elements <NE> and the total number of occurrences of elements <TNOE>.

Additional characteristics provide more detail to the measurement results for the purposes of determining structural similarity, embedding descriptions of individual elements "e" ($Elem_e^t$). in structure types "t" ($Type_t^g$) for the given group "g" (GSC^g) and each source code file (x_m, y_n) of systems X and Y, which is shown in Fig. 6.

Based on defined global $SIMstr^{GSC^g}$ and local $w_t^{GSC^g}$ similarity indexes, similarity areas are determined. A necessary condition of code similarity is that the index reaches a value close to one (≥ 0.9). If the index has a value (< 0.5), we can be fairly sure that there are major differences in the source code of the compared programs. If a value close to **1.0** is reached, a more in-depth analysis must be performed.

During this analysis it is assumed that sets of isolated lexical units in the shape of structure type names (NameOfType) and names of source file elements (NameOfTheElement) of the compared systems belonging to a specific group of source

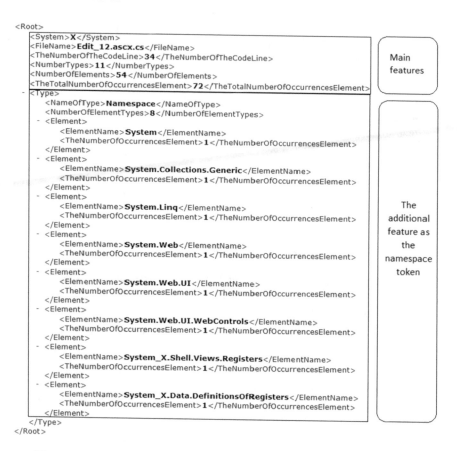

```
<Root>
  <System>X</System>
  <FileName>Edit_12.ascx.cs</FileName>
  <TheNumberOfTheCodeLine>34</TheNumberOfTheCodeLine>
  <NumberTypes>11</NumberTypes>
  <NumberOfElements>54</NumberOfElements>
  <TheTotalNumberOfOccurrencesElement>72</TheTotalNumberOfOccurrencesElement>
  <Type>
    <NameOfType>Namespace</NameOfType>
    <NumberOfElementTypes>8</NumberOfElementTypes>
    <Element>
      <ElementName>System</ElementName>
      <TheNumberOfOccurrencesElement>1</TheNumberOfOccurrencesElement>
    </Element>
    <Element>
      <ElementName>System.Collections.Generic</ElementName>
      <TheNumberOfOccurrencesElement>1</TheNumberOfOccurrencesElement>
    </Element>
    <Element>
      <ElementName>System.Linq</ElementName>
      <TheNumberOfOccurrencesElement>1</TheNumberOfOccurrencesElement>
    </Element>
    <Element>
      <ElementName>System.Web</ElementName>
      <TheNumberOfOccurrencesElement>1</TheNumberOfOccurrencesElement>
    </Element>
    <Element>
      <ElementName>System.Web.UI</ElementName>
      <TheNumberOfOccurrencesElement>1</TheNumberOfOccurrencesElement>
    </Element>
    <Element>
      <ElementName>System.Web.UI.WebControls</ElementName>
      <TheNumberOfOccurrencesElement>1</TheNumberOfOccurrencesElement>
    </Element>
    <Element>
      <ElementName>System_X.Shell.Views.Registers</ElementName>
      <TheNumberOfOccurrencesElement>1</TheNumberOfOccurrencesElement>
    </Element>
    <Element>
      <ElementName>System_X.Data.DefinitionsOfRegisters</ElementName>
      <TheNumberOfOccurrencesElement>1</TheNumberOfOccurrencesElement>
    </Element>
  </Type>
</Root>
```

Main features

The additional feature as the namespace token

Fig. 5. Sample fragment of a source code description using a name space token

⊞ (RootType)	
System	string
FileName	string
TheNumberOfTheCodeLine	int
NumberTypes	int
NumberOfElements	int
TheTotalNumberOfOccurrencesOfElement	int
Type	(TypeType)

Root

⊞ (TypeType)	
NameOfType	string
NumberOfElementTypes	int
Element	[1..*] (ElementType)

$Token_t$

Attributes of structural similarity

THE STRUCTURE OF THE DESCRIPTION OF THE SOURCE CODE FILE USING THE TOKENS

Fig. 6. The xsd structure of a source code file description using a token

code files GSC^g are sets of concepts (elements of a dictionary kept in the database), such as: *property names, method names, class names,* etc.

In the next step, concept elements (dictionary) belonging to objects in the analysed systems are compared to a search for identical names, which is done with the query below.

```
SELECT DISTINCT S_X_ASPX_CS.TheNumberOfTypes, S_X_ASPX_CS.NumberOfElement,
S_Y_ASPX_CS.NumberOfElement, S_X_ASPX_CS.TheNumberOfTheCodeLine,
S_X_ASPX_CS.NazwaPliku, S_Y_ASPX_CS.NazwaPliku,
S_Y_ASPX_CS.TheNumberOfTheCodeLine
FROM S_X_ASPX_CS INNER JOIN S_Y_ASPX_CS ON S_X_ASPX_CS.[TheNameOfTheType] =
S_Y_ASPX_CS.[TheNameOfTheType]
WHERE ((S_X_ASPX_CS.TheNumberOfTypes) = [S_Y_ASPX_CS].[TheNumberOfTypes])
AND ((S_X_ASPX_CS.NumberOfElement) Between
(0.95*[S_Y_ASPX_CS].[NumberOfElement]) And
(1.05*[S_Y_ASPX_CS].[NumberOfElement])) AND
((S_X_ASPX_CS.Filename)=[S_Y_ASPX_CS].[Filename]) AND
((S_X_ASPX_CS.NameOfTheElement) = [S_Y_ASPX_CS].NameOfTheElement]));
```

Under the proposed Methodology of software similarity study – M3S, an additional operation is performed on the above-mentioned dictionaries. The operation involves searching for "synonyms" (using the *Levenshtein metric*), i.e. as defined by the dictionary concept methodology – for terms where disruptions were introduced by switching the order of characters or adding a constant prefix or suffix to the dictionary term. As a result of this operation, both identical and similar term names are detected.

Having the "similar source files" found using the query, we can manually determine the SLINE metric value.

For comparing important fragments of source code of systems X and Y, contained in files x_i and y_j:

$$SIM_{Line}(x_i, y_j) = \frac{2 \cdot l_w^T(x_i, y_j)}{l_w^{IND}(x_i) + l_w^{IND}(y_j)}$$

where:
$l_w^T(x_i, y_j)$ - means the number of lines in identical sequences of the compared source code files of systems with the assumed length α. contained in files with indexes x_i, y_j, respectively;
$l_w^{IND}(x_i), l_w^{IND}(y_j)$ - mean the total numbers of significant lines of code, constituting the result of the creative work of the programmer.

4   Methodology of Software Similarity Study - M3S

the proposed M3S (Methodology of Software Similarity Study), in its first phase, source code files of systems X and Y are subjected to syntax-controlled analysers, i.e. they are transformed into descriptions in e form of sets of tokens, recorded as .xml files. As with the source code files, these descriptions form separable GSC groups. Next, based on the descriptions created (contained in XML files), a diagram of the database is created and data are saved in it. This provides a broad range of options to study system similarity by forming the right queries and using the BI environment for data analysis. Through mining of the processed data, values of the developed metrics are determined,

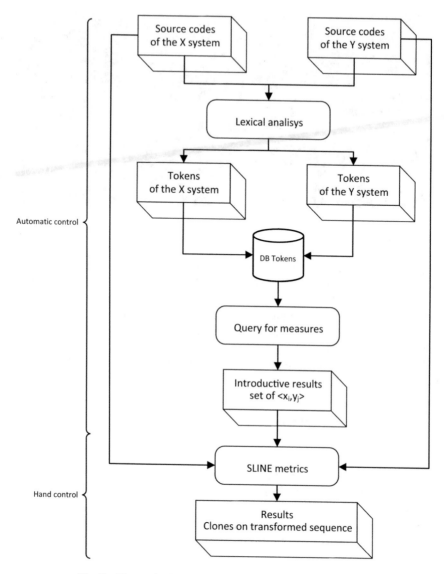

Fig. 7. The methodology of software similarity study (M3S)

and in particular areas with highly similar structural characteristics are identified, i.e. with identical names, with a specific similarity or identical signatures. Additionally, pairs of files for manual analysis are identified, i.e. analysis using text comparison tools (Fig. 7).

As a result of this operation, similarity expressed using the SIM_{Line} is determined, which concerns identical lines of source code in the analyzed source files of the studied systems. For the purposes of the developed methodology, Software Similarity Study Tools 3ST were specified, comprising:

- Syntax-controlled analyzers /authors' own software/;
- Database /MS Access/;
- Data analysis tools /Power BI/;
- Text comparison tools from the platform: Jazz Team Server Change and configuration management.

5 3ST Application

The 3ST environment developed for the purposes of the M3S was used to perform a software similarity study in court expert opinion preparation processes. Within one of them, the similarity between system X and 4 consecutive versions of system Y were analyzed. Selected characteristics of these systems are shown in the table below (Table 3).

Table 3. Data processed for the purposes of court expert opinion.

The number of types	Number of element	The number of the code line	The number of occurrences of element	Number of files	System
5694	90027	97976	296752	705	X
3382	42732	47391	102416	1969	Y
2215	35208	23266	111044		Y_1
2268	35392	23243	110458		Y_2
1316	20062	21925	53582		Y_3
1970	31829	32409	79791		Y_4
Measures of similarity of systems					
0,7453	0,6438	0,6520	0,5131	0,5273	X2Y
0,3755	0,3645	0,3657	0,3059	0	$X2Y_3$
0,5141	0,5224	0,4971	0,4238	0	$X2Y_4$

Based on the assessments conducted, it was found that in the business logic layer, the code similarity index is approximately 3.5%, and in the presentation layer it is 0.5% – for ASCX files containing definitions and interface own control support, and 3.33% – for ASPX files containing system definitions and interface page support. The interface element identifier similarity index in ASPX files is about 5%, while for ASCX own control definition files it is 0.5%. The highest level of similarity is exhibited by definitions of application interface elements in HTML – 100% due to the use of all interface element definitions from HTML in both systems (Fig. 8).

Structural similarity within code file groups for the following measures:

<The number of types>; <Number of Element>; <The number of the Code Line>; <Number of files> is shown in the figure below (Fig. 9).

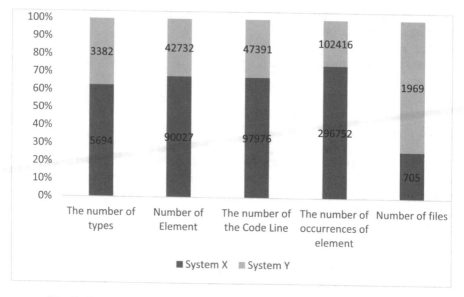

Fig. 8. Results of comparing global structural similarity of systems X and Y

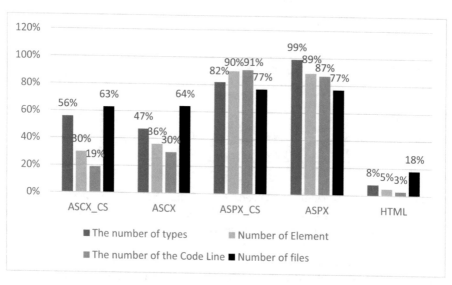

Fig. 9. Results of comparing structural similarity within code file groups

6 Discussion and Related Work

Software metrics can be classified into three categories [9]: product metrics, process metrics, and project metrics. Product metrics describe the characteristics of the product such as size, complexity, design features, performance, and quality level. Process metrics can be used to improve software development and maintenance. Project metrics describe the project characteristics and execution in terms of the number of software developers, the staffing pattern over the life cycle of the software, cost, schedule, and productivity.

The use of software metrics in software engineering field mainly applies to controlling software quality, efficiency and software production processes, or its testing. For these purposes, many attempts to use software metrics have been developed and widely presented in the literature, such as [9–11], and many others. In the book [9], an attempt to provide complete coverage of the type of metrics and models in software quality engineering was provided. The work [9] categorizes and covers four types of models: (1) quality management models; (2) software reliability and projection models; (3) complexity metrics and models; and (4) customer-view metrics, measurements, and models.

Various research on finding software similarities has been performed, most of which focused on detecting program plagiarism [5–7] and clones [20]. The usual approach extracts several metric values (or attributes) characterizing the target programs and then compares those values. Ottenstein in the work [12] used Halstead's meter measurement [11] of the target program files to compare similarity. There are other approaches which use a set of metric values to characterize source programs [13–15].

Also, structural information has been employed to increase precision of comparison [16, 17]. In order to improve both precision and efficiency, abstracted text sequences (token sequences) can be employed for comparison [7, 18, 19]. Source code texts are translated into token sequences representing programs structures, and the longest common subsequence algorithm is applied to obtain matching. The similarity metric values calculated by comparing the values of the metrics do not show the ratio of similar codes to the remaining code. These systems are aimed mainly at finding similar software code in the education environment and scalability of those evaluation methods to large software system is not known.

In the evolution of long-live software systems many different version are elaborated and delivered. Several distinct versions may be unified later and merged into another version. Knowing development relations and similarity among such systems is a key to efficient development of new systems. In [4], a similarity metric between two sets of source code files based on the correspondence of overall source code lines is proposed.

The study of the similarity between documents is presented by Broder [1]. In this approach, a set of sequences of fixed-length tokens is taken from documents. Then, two sets of X and Y are obtained for each document to calculate their intersection. Similarity is defined as $(|X| \cap |Y|)/(|X| \cup |Y|)$. This approach is suitable for efficiently calculating the similarity of a large collection of documents, such as web documents around the world. This is probably the wrong approach to calculating subjective similarity metrics for source code files.

Manber [2] developed a tool to identify similar files in large systems. This tool uses a set of keywords and extracts substring starting from these keywords as fingerprints. The set of fingerprints X of the target file is coded and compared to the set of fingerprints of the query file Y. Similarity is defined as $|X \cap Y| / |X|$. This approach works very well for both source files and document files and fits the exploration of similar files in a large system. However, he is fragile in the selection of key words.

In addition, it would be too sensitive to minor modifications to the source program files, such as changing identifiers and inserting comments.

The Broder's and Manber's methods differ significantly from the methods developed and presented in this paper, because they do not make comparisons on raw and general text sequences, but rather on sampled text sequences. The sampling approach would yield high output, but the resulting similarity would be less significant than our approach to comparing the entire text.

7 Conclusion (and Ongoing Work)

The approach presented in the paper was effective and was successfully applied by the authors to prepare court expert opinions. Due to publishing limitations, the paper only presents selected elements of the method developed, and in particular the results of the software architecture similarity tests performed were omitted.

The methodology presented in the paper was verified using complex examples of actual systems involved in court disputes. Its effectiveness was confirmed in the proceedings. The methodology provides unbiased data for determining the identity of source code of the analyzed systems, and enables detecting identical sequences and determining: global, group, local and type-based similarity indexes of system software. In a further paper, we intend to expand the description of the methodology with an assessment whether system Y was built on the basis of system X architecture, and in particular, what the level of architectonic similarity between the analyzed system was.

References

1. Broder, A.Z.: On the resemblance and containment of documents. In: Proceedings of Compression and Complexity of Sequences, pp. 21–29 (1998)
2. Manber, U.: Finding similar files in a large file system. In: USENIX Winter 1994 Technical Conference, San Francisco, CA, USA, pp. 1–10 (1994)
3. Kemerer, C.F., Slaughter, S.: An empirical approach to studying software evolution. IEEE Trans. Softw. Eng. **25**, 493–509 (1999)
4. Yamamoto, T., Matsushita, M., Kamiya, T., Inoue, K.: Measuring similarity of large software systems based on source code correspondence. In: Proceedings of 6th International Conference, PROFES 2005, Oulu, Finland, 13–15 June 2005
5. Wise, M.J.: YAP3: improved detection of similarities in computer program and other texts. In: SIGCSEB: SIGCSE Bulletin (ACM Special Interest Group on Computer Science Education) 28 (1996)
6. Schleimer, S., Wilkerson, D., Aiken, A.: Winnowing: local algorithms for document fingerprinting. In: Proceedings of the ACM SIGMOD International Conference on Management of Data, pp. 76–85 (2003)
7. Kamiya, T., Kusumoto, S., Inoue, K.: CCFinder: a multilinguistic token-based code clone detection system for large scale source code. IEEE Trans. Softw. Eng. **28**, 654–670 (2002)
8. Levenshtein, V.I.: Binary codes capable of correcting deletions, insertions, and reversals. Sov. Phys. Dokl. **10**(8), 707–710 (1966)

9. Kan, S.H.: Metrics and Models in Software Quality Engineering. Addison-Wesley Longman Publishing, Boston (2002)
10. Preece, J., Rombach, H.D.: A taxonomy for combining software engineering and human-computer interaction measurement approaches: towards a common framework. Int. J. Hum.-Comput. Stud. **41**(4), 553–583 (1994)
11. Halstead, M.H.: Elements of Software Science. Elsevier, New York (1977)
12. Ottenstein, K.J.: An algorithmic approach to the detection and prevention of plagiarism. ACM SIGCSE Bull. **8**, 30–41 (1976)
13. Clements, P., Northrop, L.: Software Product Lines: Practices and Patterns. Addison Wesley, Boston (2001)
14. Donaldson, J.L., Lancaster, A.M., Sposato, P.H.: A plagiarism detection system. In: ACM SIGCSE Bulletin (12th SIGSCE Technical Symposium on Computer Science Education), vol. 13, pp. 21–25 (1981)
15. Grier, S.: A tool that detects plagiarism in pascal programs. In: ACM SIGCSE Bulletin (12th SIGSCE Technical Symposium on Computer Science Education), vol. 13, pp. 15–20 (1981)
16. Jankowitz, H.T.: Detecting plagiarism in student Pascal programs. Comput. J. **31**, 1–8 (1988)
17. Verco, K.L., Wise, M.J.: Software for detecting suspected plagiarism: comparing structure and attribute-counting systems. In: Rosenberg, J. (ed.) Proceedings of 1st Australian Conference on Computer Science Education, Sydney, Australia, pp. 86–95 (1996)
18. Gusfield, D.: Algorithms on Strings, Trees, and Sequences. Computer Science and Computational Biology. Cambridge University Press, Cambridge (1997)
19. Whale, G.: Identification of program similarity in large populations. Comput. J. **33**, 140–146 (1990)
20. Koschke R.: Survey of research on software clones. In: Duplication, Redundancy, and Similarity in Software. Dagstuhl Seminar Proceedings, Germany (2007)

How Good Is My Project? Experiences from Projecting Software Quality Using a Reference Set

Jakub Chojnacki[1], Cezary Mazurek[1], Bartosz Walter[1,2]([✉]), and Marcin Wolski[1]

[1] Poznań Supercomputing and Networking Center, Poznań, Poland
{jchojnacki,mazurek,marcin.wolski}@man.poznan.pl
[2] Faculty of Computing, Poznań University of Technology, Poznań, Poland
bartosz.walter@cs.put.poznan.pl

Abstract. Effective management of a portfolio of software projects may include ranking them with respect to various qualitative criteria. Diversity of projects in terms of their size, maturity and domain poses additional requirements for adopting a quality evaluation process. At first, to provide a convincing argumentation for project improvements, it is important to extract relevant quality features. Secondly, strong and weak aspects of a project could be identified by showing their relevance compared to other projects. In the paper we present experiences from applying a method of ranking software projects based on practical cases. We discuss the relevancy of features used for comparison and analyze various aggregation methods used for comparing projects of similar nature.

Keywords: Software quality · Software management
Project comparison

1 Introduction

Quality is one of the decisive factors on the competitive market of software systems. Instant and continuous control over the development processes increases chances to deliver a product that satisfies its users, maintainers and also developers. To provide an insight into the processes, several quality systems that embrace various processes and artifacts, have been proposed [1]. Early detected anomalies provide empirical evidence as an input for taking corrective actions.

Increasing complexity of software systems has influenced on rising interest in continuous software quality evaluation. In particular, this observation applies to the software that is developed globally, uses third-party components, operates directly on hardware devices, or is deployed in cloud environment [2,3]. Furthermore, delivering actionable advice on quality, based on automated tools and methods, is still a challenge [4]. It is also the case for generating results that help or facilitate to manage a software project. There are some external factors,

P. Kosiuczenko and Z. Zieliński (Eds.): KKIO 2018, AISC 830, pp. 192–207, 2019.
https://doi.org/10.1007/978-3-319-99617-2_13

beyond the scope of the software product, that could also affect the software process in organizations [5].

The promising method that help to evaluate the status of a project is presenting how its relative quality can be compared to the quality of other projects which are used as a reference set. This approach does not rely on a direct pairwise comparison, but rather aims at creating a proper means that would help to understand the similarities and differences between the projects, and evaluate them by analogy.

In an organization with a large software portfolio, the reference set may include projects that apply the same or similar development practices and qualitative criteria. However, the obtained results of evaluation could be biased, as the background projects would be too similar (they are developed by the same teams, with similar habits and adopting similar processes). There is a need for an approach that combines the diversity of the benchmark sample with similarity of selected features.

In this paper we present our experiences on evaluating the quality of a software project by projecting it onto a reference set of other projects. The reference set is diverse, but also displays similarity to the subject project with respect to a selected feature. That facilitates more precise identification of weak and strong points, and consequently planning adequate improvement actions.

In Sect. 2, we shortly present GÉANT organization and GN-QM quality framework. Next, we focus on presenting the method of quality evaluation: in Sect. 3 we describe the construction of a reference set, and in Sect. 4 we present results of an exemplary evaluation. Later, in Sect. 5 we summarize the literature concerning the topic. Finally, in Sect. 6, we conclude and mark out prospective developments.

2 GN-QM – A GÉANT Software Evaluation Framework

GÉANT is a research and innovation organization built upon a federation of NRENs - operators of national communication networks for science and education. The GÉANT product portfolio includes a variety of software projects of diverse maturity i.e. research prototypes, pilot solutions and fully operational application, size and target domains. The innovative nature of the developed solutions and characteristics of NRENs software developers and users has been recognized as an important factor affecting the methods of managing the product portfolio [6].

The product management in GÉANT is based on ITIL and a set of supporting processes. For example, it includes a validation and testing process that supports the product's transition to production. The process ensures that products deployed in the production infrastructure are of highest quality and meet commonly approved standards [7].

The GÉANT software quality evaluation framework (GN-QM) [8] has been proposed and added to the testing process as a tool presenting objectively how the software system is developed and maintained with respect to selected quality factors. The assessment combines the product and process perspectives towards the objective software evaluation based on metrics and measurements.

GN-QM originates from classical software quality models, but also identifies a perspective specific to innovative, research-oriented projects, and applies the specific characteristics of the networking environment. The GN-QM framework supports the maturity assessment in software products based on recognized standards, and supplied by evidence collected from the analyzed processes. It inherits the structure and several individual characteristics ISO 9216 model [9], combining selected features from other models used in open source and commercial domains, but also supplements them with items specific to the innovative systems with a limited number of users.

2.1 Data Collector

To accelerate data processing, we developed *Data Collector* – a tool for automated collecting, aggregating and reporting data relevant for a software system evaluation. The tool resembles a multi-step graphical wizard, which helps even non-experienced users to collect necessary data from the various sources and repositories: currently these are an issue tracker (JIRA), a version control system (Git/SVN) and source code analyzers (SonarQube). Moreover, it allows to enter the manually acquired data, like that coming from software audits, test reports or end-user surveys.

Architecture of the Data Collector is presented in Fig. 1. The tool queries the selected data sources to calculate values of metrics defined in the configuration file. Then, it aggregates the metrics at two levels: subcharacteristics, and further high-level characteristics. Based on the calculated scores, the Data Collector

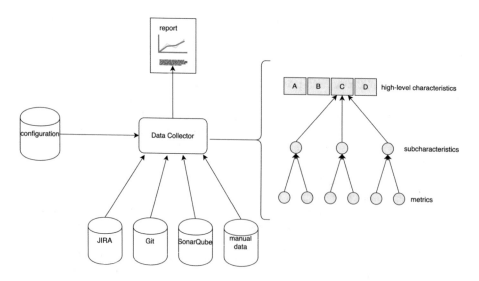

Fig. 1. Architecture of the Data Collector

generates a report that includes diagnosis of the present status and recommends actions that could improve the results.

Specifically, the tool calculates four high-level characteristics defined in GN-QM: A-Functionality, B-Maintainability, C-Marketability, and D-Reliability, as values from range 0 to 1, where 0 is the minimum, and 1 - the maximum value; the score is monotonic, i.e., a higher value indicates a more desired value. Each high-level characteristic can be evaluated as *weak* (value < 0.4), *average* (0.4 ≤ value < 0.6) or *good* (value ≥ 0.6). The values are calculated as a weighted mean of the underlying values (subcharacteristics or raw metrics), which are also subject to the same calculating procedure and interpretation (Fig. 2).

Fig. 2. An example of software quality evaluation in GN-QM

The tool has already left the prototype phase and has been used to evaluate the quality of selected GÉANT systems. In order to calibrate the metrics and perform additional tool validation, we performed an initial analysis on a number of open source projects hosted at the Apache Foundation[1].

---

[1] http://projects.apache.org.

3   Evaluation Schema

3.1   A Reference Set of Projects

To put a project in a proper perspective, we need a reference set of projects that could serve as a benchmark. The reference set needs to be curated, i.e., include projects that meet certain criteria, e.g., representativeness or diversity, and have defined procedures for including and removing the projects from it.

In our case we adopted the following criteria to define the reference set:

- *Availability* – projects should be freely available in both the source code and associated artifacts (issues, release notes, repository etc.);
- *Diversity* – projects in the reference set should be diverse with respect to selected features;
- *Completeness of data* – the scores in GN-QM should be complete for each projects and rely on several available data sources, i.e., Jira, GIT and Sonar.

In conformance to these criteria, we performed maturity evaluation for a number of open source projects hosted by Apache Foundation. To collect data, we created a dedicated tool that retrieves, information from various data sources, and then reports the maturity levels defined in the GN-QM framework. In this case, the maturity level is expressed by only three high-level quality characteristics of GN-QM: Maintainability, Marketability and Reliability. Functionality was excluded from analysis, as it relies on metrics dependent on data which was not available for all projects.

In Table 1 we provide a list of projects currently included in the reference set, along with basic descriptive data.

Although several of these projects come from Apache Commons, every component is developed independently, by a dedicated team with its own conventions. Thus, at least the minimal diversity of the set is preserved.

3.2   Contextual Features

Due to the diversity and variety of projects included in the reference set, presenting a subject project in a perspective of the entire reference set could not bring useful information concerning the relative quality of the project. For example, comparing a small project against a number of other projects of different sizes could lead to formulating inadequate recommendations. As a result, using a highly diverse reference set as a background may not be an optimal choice. We observe a trade-off between two contradictory requirements: on one hand, diversity of a set of projects is a deserved feature [10] but, on the other hand, the compared projects should be of similar nature, to put the results in a proper context.

Software projects could be categorized with respect to common features that define their context. Capiluppi et al. [11] identified 12 such features, e.g., project age, application domain, programming language, etc. Since not all of them are

Table 1. Reference set of projects from Apache Foundation.

Name	Description
Commons BCEL	Tool for analyzing and manipulating Java class files
Cayenne	Database mapping tool
Commons Chain	Implementation of the Chain of Responsibility pattern
Commons Collections	Common Java data structures and algorithms
Commons Configuration	Generic configuration library
Commons DBCP	Abstraction layer for interacting with a relational database
DDF	Distributed Data Framework, integration framework
Commons Digester	XML-to-Java objects mapping module
Empire-db	High level object-oriented API for RDBMS
Hama	Big Data and high-performance computing
Maven Enforcer	Control certain environmental constraints
nuxeo	Content management platform
OGNL	Object-Graph Navigation Language, Java extension
Pluto	Java Portlet implementation
Commons Pool	Implementation of Pool of Objects pattern
Rampart	Implementations of the WS-Sec specifications
Maven SCM	Common API for performing SCM operations
Commons SCXML	Generic state-machine based execution environment
Sirona	A monitoring solution for Java applications
Tika	Toolkit for detecting and extracting metadata and text
Commons VFS	Single API for different file systems

relevant or available, we decided to focus on five core features that could be extracted for all projects:

- **system size** – expressed by SLOC (source lines of code),
- **system age** – expressed by the time that passed between the first and the last activity in the project,
- **user activity** – expressed by the number of issues reported in an issue tracking system,
- **team size** – expressed by the number of people who contributed to the project since its beginning
- **development activity** – expressed by the number of commits

3.3   Stratification of Data

Each of the common contextual features allows for partitioning the reference set into smaller slices that include projects similar with respect to that feature. In this case, we decided to split each feature into four slices that include comparable number of projects in each slice. Then, we have the slices representing first (Q_1), second (Q_2), third (Q_3) and fourth (Q_4) quartile of the reference set. The quartiles are defined separately for each of the contextual features, i.e., in our case, five different partitions into slices exist.

The concept of stratification is presented in Fig. 3. A subject project (marked by a square) is placed in the third quartile with respect to the code size (depicted in horizontal axis), and the fourth quartile, according to the team size (the vertical axis). Therefore, different projects would be used as a background for presenting the subject project, depending on the contextual feature used for stratification. One of the variability factors is removed, which can also affect the recommendations inferred from the comparison. It also outlines how the choice of of contextual features affects the results; therefore, the feature must be provided along with the results.

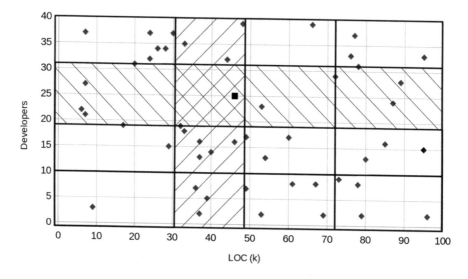

Fig. 3. Stratification of the reference set for two exemplary features.

Once a slice of the reference set is selected, we can present the subject projects' score with respect to that slice. Since the slice represents projects similar to the subject, the recommendations could be more focused and better more context-oriented than if the project were compared against the entire reference set.

There are two remarks concerning the presented approach.

By selecting a feature used for slicing the reference set, we also alter the set of specific projects included into the slices, which in turn affects the results of comparison, as the subject project is compared against other background projects. Therefore, in order to be informative, the results need to be supplemented with the information concerning the feature used for slicing.

Additionally, some aspects of the project quality, including specific scores, can be considered confidential, and should not be disclosed beyond its stakeholders, even within the same organization. Publishing a ranking of projects with respect of their quality may trigger unnecessary emotions.

To avoid that, we decided to make the results of evaluation available at two levels:

1. comprehensive information, including values of individual metrics and recommendations for improvement is delivered to the team leaders, to provide them with quantitative data and recommendations on how to manage the projects to improve its scores;
2. the relative position of the subject project with respect to the slice used as background (e.g., "the subject project is better than 40% of projects of similar size") can be published within the organization, at the discretion of the team leader.

4   Example of Application

As an example of the presented approach, we present the evaluation of three projects from the GÉANT ecosystem: System 1, System 2 and System 3, using the Apache projects as a reference set.

System 1 provides a service for creating isolated customized virtual networks, aiming at verifying novel concepts in networking and telecommunications. It is designed to support testing and development of advanced networking technologies in large-scale, dispersed environments. The system is of medium size (30k LOC), written entirely in Java, and has been actively developed since 2013 by 12 committers. Since 2014, it has been in its maintenance phase, handled by 4 developers.

System 2 is a web portal that allows students and researchers to work in a secure environment. Its primary function is helping its users to connect with each other using VPN connections. Additionally, it is aimed at reaching a broad community of users, so System 2 does not require extensive technical knowledge to handle it properly. System 2 is also a medium-sized project (30k LOC), written mostly in Java and JavaScript. The project was developed in 2014–2016 by 5 programmers.

System 3 is a Java-based web application that enables users to visualize and help in troubleshooting network issues. It manages a set of plugins, responsible for conducting measurements and collecting the results. System 3 started in 2012 and underwent intensive development in the years 2013–2016, involving 7 programmers. Nowadays, it is still maintained by 2 developers, who occasionally apply minor patches and improvements.

These projects above have been chosen with regard to the availability and completeness of the quality related data (existing in the supporting systems like Jira, GIT and SONAR), as well as to represent various application and technological domains.

In Table 2 we present results for three selected characteristics (B-Maintainability, C-Marketability and D-Reliability) of Systems 1, 2 and 3, along with the evaluation of reference projects. Next, we present and shortly discuss projections of the projects with respect to selected common features. We focus on interesting and not-so-obvious cases, to show how the results could be interpreted.

In Fig. 4 we present the evaluation of characteristic B-Maintainability, with respect to #commits as the slicing feature. We observe that System 1 belongs to the Q4, while System 2 and System 3 is assigned to the Q1. Even though the absolute score for System 1 is higher than for System 2 and 3, the relative position within its slice is comparable to the positions of remaining systems in

Table 2. Sample subject systems and the reference set of Apache projects

Name	GN-QM			Values of the slicing features				
	B	C	D	Team	Size	#k-commits	#k-issues	Lifetime
System 1	0.58	0.22	0.83	14	124	7.7	0.9	5
System 2	0.54	0.21	0.61	7	27	0.9	0.5	2
System 3	0.49	0.22	0.59	9	35	0.6	0.7	8
Cayenne	0.61	0.16	0.89	27	147	6.0	2.4	11
Commons BCEL	0.50	0.11	0.64	16	31	1.3	0.3	17
Commons Chain	0.39	0.14	0.77	15	6	0.6	0.2	14
Commons Collections	0.56	0.28	0.71	29	29	3.0	0.7	17
Commons Configuration	0.50	0.26	0.72	21	28	3.0	0.7	14
Commons DBCP	0.40	0.14	0.67	22	12	1.5	0.5	16
Commons Digester	0.41	0.17	0.75	17	13	2.0	0.2	17
Commons Pool	0.48	0.22	0.69	23	6	1.5	0.4	16
Commons SCXML	0.46	0.26	0.71	13	10	1.0	0.3	13
Commons VFS	0.48	0.25	0.71	21	23	2.0	0.7	16
DDF	0.85	0.37	0.85	101	164	9.0	3.0	5
Empire-db	0.50	0.37	0.70	4	44	1.0	0.2	10
Hama	0.61	0.40	0.69	15	40	1.5	1.0	9
Maven Enforcer	0.39	0.35	0.67	17	5	0.5	0.3	11
Maven SCM	0.65	0.28	0.68	34	50	2.0	0.8	14
Nuxeo	0.54	0.17	0.81	84	560	35.0	24.0	10
OGNL	0.74	0.11	0.67	8	14	0.6	0.3	7
Pluto	0.69	0.34	0.53	6	26	2.0	0.7	11
Rampart	0.61	0.31	0.61	4	21	1.0	0.5	11
Sirona	0.67	0.28	0.64	8	46	1.0	0.1	10
Tika	0.80	0.40	0.86	65	55	4.0	2.5	11

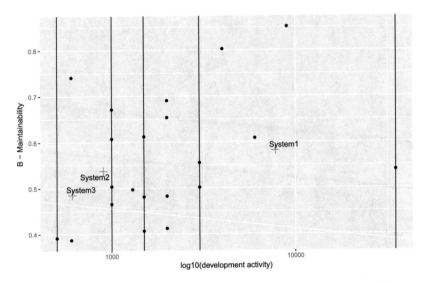

Fig. 4. Evaluation of B-Maintainability characteristic with respect to #commits

their slice. It exemplifies a case, in which the selected slicing criterion seems to have a decisive impact on the evaluation: if the subject systems were compared to the entire reference set, the results for relative comparison would be different.

Another case is depicted in Fig. 5. The quartiles extracted for the duration feature are of different width, and all three systems have been included into Q1. As a result, their relative evaluation scores are similar.

The GN-QM score for the D-reliability of the subject systems is significantly different than for the B-maintainability, which results in different relative evaluation. In Fig. 6 we present a diagram of the D characteristic with respect to #commits. System 1, included in the Q4, clearly outranks remaining subject systems, which are assigned to Q1. However, System 1 has lower rank in its slice than three other reference systems, which means that its position within the slice would not be as high, as the overall reliability score suggests.

Finally, in Fig. 7 we observe that all subject systems have been included in the Q1 slice, which is also the widest slice in the reference set. System 2 and 3 are exceeded by all comparable systems in the slice, while System 1 has the highest evaluation. In this diagram, we can see how the choice of the slicing feature could affect the relative evaluation of the subject system.

Based on the presented examples, we may draw following conclusions:

1. Missing information about the common feature used for slicing the reference hinders interpretation or even makes in impossible.
2. Slices should include at least five instances of reference projects, to provide reliable benchmark for comparison.
3. Uniform partition of the reference set produces slices with equal or similar size, which helps in interpretation of the result.

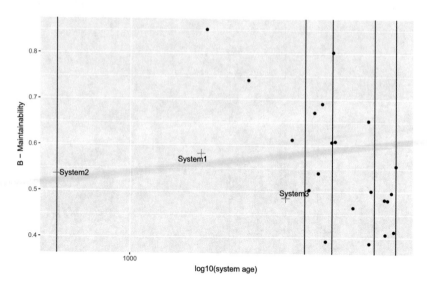

Fig. 5. Evaluation of B-Maintainability characteristic with respect to system lifetime

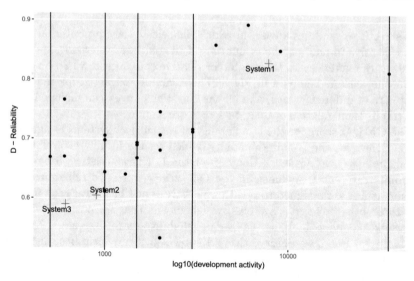

Fig. 6. Evaluation of D-Reliability characteristic with respect to #commits

4.1 Redundancy of Some Features

After analyzing the examples of use, we clearly see that some features partition the reference set into very similar slices. In that case, using more features does not make the evaluation more complete.

In Fig. 8 we present the matrix of Pearson correlation for the pairs of slicing features, calculated for the exemplary data presented in Sect. 4. Based on that,

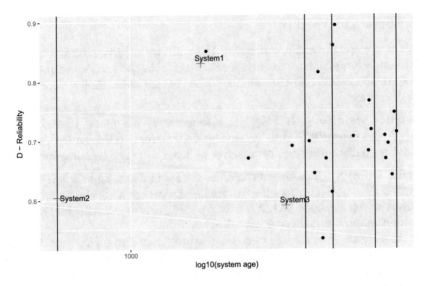

Fig. 7. Evaluation of D-Maintainability characteristic with respect to system lifetime

we notice high correlation for code size and #commits, size and #issues, and #commits and #issues. It indicates that these features actually represent similar information and could be considered as redundant. Additionally, the team size is also correlated with code size, #commits and #issues.

From all slicing features, only the system age is not correlated with other features; then, at minimum it is justified to compare the evaluations of characteristics with respect to that feature and at least one feature from the set code size, #commits, #issues, team size.

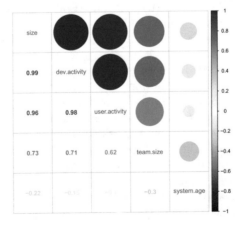

Fig. 8. Pearson correlation matrix for slicing features

4.2   Threats to Validity

In this paper we present a concept of the software quality evaluation with an initial validation against a reference set of projects. Although the preliminary results are promising, there is a number of issues that could limit its applicability in other context.

Construct validity – GeantQM is a framework customized to the needs of GÉANT, and captures the quality dimensions relevant to this entity. As such, it may not properly reflect the perspective of other software organizations.

Internal validity – The size of the reference set in its current form is relatively small, and does not allow for effective slicing nor generalizing the findings. Additionally, the reference set includes only Apache projects, which could also make it too uniform and not diverse enough.

Moreover, the correlation between features used for slicing was identified only with respect to the selected GN-QM characteristics, and only with a small sample of projects. That does not allow for making conclusions concerning the dependencies among the contextual features we used.

External validity – At the moment the presented concept was exemplified and should be validated on a larger sample of projects before putting into practice. There is also a need of comparing the result with an external perception of quality, to validate the usefulness of the approach.

5   Related Work

5.1   Ranking Software Projects

A typical approach for ranking software projects is based on benchmarks. A benchmark is usually defined based on repositories of metrics and their corresponding thresholds [12–14], to enable comparison of the subject system with others, with respect to code quality and maintainability. As noted in [15], benchmarks that rely on commonly used metrics should be tailored to a particular context.

An alternative approach has been recognized by software industry for systematic comparison and evaluation of open source projects. It is a vital method in supporting industry *buy vs. build* decision making processes. In response to that, several methodologies for conducting the assessment have been proposed, aiming at selecting a subset of suitable FLOSS projects from a number of available options. Notable examples include OpenBRR OMM [16] or QSOS [17].

The software quality framework GN-QM, used as a base for the evaluation method proposed in this paper, relies on metrics and measures to derive information about various quality aspects. However, in contrast to the typical software benchmarking mentioned in the paragraph above, our method presents not only a quality assessment score for a project, but also emphasizes the potentially strong and weak aspects of a subject system, giving as an evidence the references to other similar projects.

5.2   Features and Attributes of Software Projects

Software applications and projects can be categorized according to different contextual attributes, e.g., team size, popularity, project activity, programming language(s), project domain (e.g., application, database, framework, library) etc. The top-down approach can base the taxonomy on the already defined attributes, e.g., mentioned by ISBSG[2], in the previous research work by Baishakhi et al. [18], and on "project dimensions" [10]. Various contextual variables that affect the development of software systems are considered in [19]. The authors identified thresholds of six object-oriented software metrics using three context factors: application domain, software types and system size.

The trade-off between diversity and representativeness of project sample with respect to the context has been brought up in [10]. The authors proposed a vocabulary and techniques for measuring how well a given sample covers a population of projects. The vocabulary defines a universe as a large set of projects and dimensions for describing each of the project inside a universe, e.g., total lines of code, number of developers, main programming language and others.

A study of programming languages and code quality in GitHub has been presented in [18]. The authors characterized the projects by its complexity and the variance along various dimensions of programming language, language type, usage domain, amount of code, sizes of commits, etc.

The recent studies on GitHub repositories identified another dimension of a software project structure by showing the relevance of its software community in terms of experiences and contributions [20]. These results has not been applied in our method, but it could be added in the future, as an extension to the currently used contextual features [11].

6   Summary

Based on the experiences from evaluating GÉANT projects presented in the paper we can formulate the following conclusions.

First, the presented approach introduces a well-adjusted method of evaluating the project's quality. It relies on projection of the subject system on a set of other projects of similar nature, with respect to at least one feature. That not only gives a general understanding of how the subject project performs, compared to other projects, but can also be performed quickly without making several pairwise comparisons between projects.

Secondly, diversity of the reference set helps to generate slices of similar projects in a more balanced way. It also allows for reducing bias related with projecting a subject item onto an overly-focused sample of other projects. Therefore, the reference set should be curated and not include a random sample of projects.

However, the use of an entire reference set as a background for comparison also has its drawbacks. It could make the recommendations not specific enough to

---

[2] http://isbsg.org/software-project-data/.

the context of the subject project, which could limit their applicability. Then, the subject project should be evaluated with respect to a group of similar projects, selected from the reference set. That allows for a more focused and contextualized quality evaluation.

Another remark concerns the size of the reference set. We expect that a sufficiently large set could be partitioned with respect to several or even all subject features. That would produce uniform slices of projects, similar with respect to several criteria. On the other hand, the extensive effort related to the data acquisition may overrun the benefits. Additionally, since some features used for slicing generate highly correlated partitions of the reference set, it is likely that this approach would not improve the evaluation of relative quality of projects.

The proposed method is in the early stage of development. The first feedback from the software teams developing the evaluated systems, confirmed that results are promising and thereby encourage to carry on the further works. As a next step, a pilot solution with actionable advices for practitioners is going to be deployed and run against a wider collection of GÉANT software projects. In the future, we plan to focus on extending the reference set and defining more comprehensive rules for including new projects.

Acknowledgements. GÉANT Limited on behalf of the GN4-2 project. The research leading to these results has received funding from the European Union's Horizon 2020 research and innovation programme under Grant Agreement No. 731122 (GN4-2).

This work is financed from financial resources for science in the years 2016–2018 granted for the realization of the international project co-financed by Polish Ministry of Science and Higher Education.

References

1. Ferenc, R., Hegedűs, P., Gyimóthy, T.: Software product quality models. In: Mens, T., Serebrenik, A., Cleve, A. (eds.) Evolving Software Systems, pp. 65–100. Springer, Heidelberg (2014)
2. Spinellis, D.: Software reliability redux. IEEE Softw. **34**(4), 4–7 (2017)
3. Kazman, R.: Software engineering. Computer **50**(7), 10–11 (2017)
4. Washizaki, H.: Pitfalls and countermeasures in software quality measurements and evaluations. In: Advances in Computers, vol. 107, pp. 1–22. Elsevier (2017)
5. De Camargo, K., Curcio, A.M., Reinehr, S., Paludo, M.: An analysis of the factors determining software product quality: a comparative study. Comput. Stand. Interfaces **48**, 10–18 (2016)
6. Bilicki, V., Golub, I., Vuletic, P., Wolski, M.: Failure and success - how to move toward successful software development in Networking. In: Terena Networking Conference (2014)
7. Wolski, M., Golub, I., Frankowski, G., Radulovic, A., Berus, P., Medard, S., Kupiński, S., Apfel, T., Nowak, T., Visconti, S., Smud, I., Mazar, B., Marovic, B., Promiński, P.: Deliverable D8.1 service validation and testing process. Technical report 691567 (2016)

8. Wolski, M., Walter, B., Kupiński, S., Chojnacki, J.: Software quality model for a research-driven organization - an experience report. J. Softw. Evol. Process **30**(5), e1911 (2017)
9. Wolski, M., Walter, B., Kupinski, S., Prominski, P., Golub, I.: GN4-1 white paper: supporting the service validation and testing process in the GÉANT project. Technical report 691567 (2016)
10. Nagappan, M., Zimmermann, T., Bird, C.: Diversity in software engineering research. In: Proceedings of 2013 9th Joint Meeting of the European Software Engineering Conference and the ACM SIGSOFT Symposium on the Foundations of Software Engineering, ESEC/FSE 2013, pp. 466–476 (2013)
11. Capiluppi, A., Lago, P., Morisio, M.: Characteristics of open source projects. In: Proceedings of 7th European Conference on Software Maintenance and Reengineering (CSMR 2003), Benevento, Italy, 26–28 March 2003, p. 317 (2003)
12. Baggen, R., Correia, J.P., Schill, K., Visser, J.: Standardized code quality benchmarking for improving software maintainability. Softw. Qual. J. **20**(2), 287–307 (2012)
13. Alves, T.L., Ypma, C., Visser, J.: Deriving metric thresholds from benchmark data. In: IEEE International Conference on Software Maintenance, ICSM (2010)
14. Yamashita, A.: Experiences from performing software quality evaluations via combining benchmark-based metrics analysis, software visualization, and expert assessment. In: Proceedings of 2015 IEEE 31st International Conference on Software Maintenance and Evolution, ICSME 2015, October 2015, pp. 421–428 (2015)
15. Zhang, F., Mockus, A., Zou, Y., Khomh, F., Hassan, A.E.: How does context affect the distribution of software maintainability metrics? In: IEEE International Conference on Software Maintenance, ICSM, pp. 350–359 (2013)
16. Petrinja, E., Nambakam, R., Sillitti, A.: Introducing the OpenSource maturity model. In: 2009 ICSE Workshop on Emerging Trends in Free/Libre/Open Source Software Research and Development, pp. 37–41, May 2009
17. Deprez, J.C., Alexandre, S.: Comparing assessment methodologies for free/open source software: OpenBRR and QSOS. In: Jedlitschka, A., Salo, O. (eds.) Product-Focused Software Process Improvement. Lecture Notes in Computer Science (including subseries Lecture Notes in Artificial Intelligence and Lecture Notes in Bioinformatics) (LNCS), vol. 5089, pp. 189–203. Springer, Heidelberg (2008)
18. Ray, B., Posnett, D., Filkov, V., Devanbu, P.T.: A large scale study of programming languages and code quality in GitHub categories and subject descriptors. In: FSE 2014 (2014)
19. Ferreira, K.A.M., Bigonha, A.S., Bigonha, R.S., Mendes, L.F.O., Almeida, H.C.: Identifying thresholds for object-oriented software metrics. J. Syst. Softw. **85**(2), 244–257 (2012)
20. Onoue, S., Hata, H., Monden, A., Matsumoto, K.: Investigating and projecting population structures in open source software projects: a case study of projects in GitHub. IEICE Trans. Inf. Syst. **E99D**(5), 1304–1315 (2016)

Author Index

© Springer Nature Switzerland AG 2019
P. Kosiuczenko and Z. Zieliński (Eds.): KKIO 2018, AISC 830, p. 209, 2019.
https://doi.org/10.1007/978-3-319-99617-2

Printed in the United States
By Bookmasters